Hollow and Home

Hollow and Home

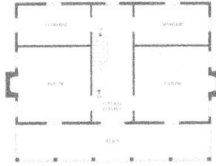

A History of Self and Place

E. FRED CARLISLE

WEST VIRGINIA UNIVERSITY PRESS
MORGANTOWN 2017

Copyright 2017 West Virginia University Press

First edition published 2017 by West Virginia University Press

ISBN
CL: 978-1-943665-81-5
PB: 978-1-943665-82-2
EPUB: 978-1-943665-83-9
PDF: 978-1-943665-84-6

Library of Congress Cataloging-in-Publication Data
is available from the Library of Congress.

Book and cover design by Than Saffel / WVU Press

For

Beth

My unexpected treasure

And for

My dear daughters

Lindy, Becky, Ginna, and Jana

Contents

LIST OF PHOTOGRAPHS AND ILLUSTRATIONS

Acknowledgments

I have benefited enormously, over the years, from the advice and support of my "constant" readers: Bob Siegle, Tom Gardner, Beth Obenshain, Edward Weisband, Jana Carlisle, and Lloyd Gardner. I am also grateful to others who have read early drafts of this and my related publications: Mary Bishop, Ralph Byers, Mark Embree, Nancy Grayson, John C. Inscoe, Tom McCormick, Toots and Scott Obenshain, and Ray Smoot.

I greatly appreciate the time, intelligence, and insight of the people I interviewed over time and who have contributed significantly to *Hollow and Home*. A list would be long, but since I identify virtually all of them in the text, I will not name them here. I cherish the memory of those who have since died. I am indebted to the late Barbara Carlisle who led me to Clover Hollow and who for years was my best reader.

I value the contributions of many people from my hometown, Delaware, Ohio. I appreciate, especially, the interest and support, for more than ten years now, of my Delaware friend Jack Hilborn. In relevant chapters, I name the others who have contributed to this book.

For photographs, archival assistance, architectural information and drawings, and general guidance, I appreciate the assistance of Emily Gattozi and Carol Holliger, Ohio Wesleyan University Library; Paul Monks, Delaware County Historical Society; Chris Hudson, Shelter Alternatives in Blacksburg, Virginia; and Bob Siegle for image and computer assistance.

I appreciate the advice of Heather Lundine, my first editor at the WVU Press. I benefited a great deal from the recommendations by the anonymous readers for the Press. I greatly value the interest, support, and editorial advice of the Press director, Derek Krissoff, from his enthusiastic reply to my initial proposal, through final editing, to publication.

The Place Is the Thing

To be at all—to exist in any way—is to be somewhere, and to be somewhere is to be in some kind of place. Place is as requisite as the air we breathe, the ground on which we stand, the bodies we have. We are surrounded by places. We walk over and through them. We live in places, relate to others in them, die in them. Nothing we do is unplaced. How could it be otherwise?

—Edward S. Casey, *The Fate of Place*

1.

James Melville Cox and Brookside Farm

"I'm looking for Fred Carlisle." I was sitting at my desk, reading some university policy or other dull document, and then: "I am looking for Fred Carlisle." I knew that deep, forceful, gravelly voice. I'd not heard it in decades, but I knew it. Jim Cox. James Melville Cox—my esteemed graduate professor at Indiana University. He was standing in my outer office addressing my secretary. I couldn't get there fast enough. "Jim! My God! How are you?" He smiled and grasped my hand. "I've thought of stopping by several times before; today I did." I drew him into my office that day in 1993. The renewal of our long suspended friendship began. It lasted until his death in 2012.

I once knew that Jim had grown up in Virginia and still owned his 1870s family farm, near Independence, Virginia. But I'd forgotten. We'd not talked or corresponded since my first years at Michigan State. So I did not know he had moved back to his home place in Grayson County after his retirement from Dartmouth at age sixty. Except for brief summer visits almost every year, he had been away from Brookside Farm for forty-three years.

I remember well Jim's classes at Indiana University. A deep sense of the comic and absurd—perhaps of the perverse—seemed always to drive his intelligence and his talk. He loved talking about Poe's "Imp of the Perverse" and Mark Twain's humor. In class, his eyes flashed knowingly; a wry smile flickered repeatedly across his face. He leaned over the table, shoulders hunched, chin down, moving his head from side to side as he talked. He spoke rapidly, insistently, and passionately from deep in his throat—one brilliant insight after another. In those graduate school days, we marveled and wondered, unsure sometimes whether this was amazing genius or a step or two short of madness. And when he laughed, it was at once ebullient and sardonic; it was a laugh of discovery, excitement, difference. His relentless intelligence worked on texts and ideas as no one else's did. His sense of irony exposed contradictions no one else could see. "Jim Cox has the most original mind I've ever encountered," Russ Nye, himself a distinguished literary historian at Michigan State, once told me.

I was inspired by Jim's intelligence and originality. But I did not fully appreciate his cultivated spirit of perversity or his deeply ironic vision of both North and South until we resumed our friendship. Nor did I understand the source of his ironic vision.

At Indiana, he taught me a lesson about sentimentality and nostalgia I have not forgotten in over fifty years. The subject was *Huckleberry Finn* and all those southern gestures and customs Huck observes traveling down river. Sentimentality masks cruelty. That's the lesson. Nostalgia obscures hardship and suffering. Later, Jim extended that to those still possessed by the Civil War—those who celebrate it and seek a finest hour in that vicious struggle. "It's awful," he said. "You don't want to see the war that way." They forget at the same time they remember. Their nostalgia masks the terrible destruction, suffering, and death while letting them forget—or at least suppress—the real war and all those deaths.

Over the next several years, we talked at length about his family history, the farm, the North and the South, his sense of place, and his relationship to the past. In Virginia after we'd resumed our friendship, we sat at the round table in his kitchen, opposite an old gas stove with cooking tools and pans hanging on the wall above it and where a row of sand-colored cabinets lined the wall. Jim and his wife, Marguerite, spent many of their indoor, waking

Image 1. James M. Cox. Image by author.

hours there. I don't remember a TV or radio. Every time I visited, I walked past the handsome facade of the 1877 farmhouse, around to the far side, and entered through the kitchen door into their true living room. I never entered or left through the front door of the house.

We talked for hours at a time. Jim's hair had turned white, but his complexion reminded me of his light red hair and freckles when I first knew him. He relaxed in his chair and drank coffee as he talked and gestured. He was thoughtful and reflective, not quite the intense and driven man he was at Indiana. His smile and laugh, however, had not changed much. They still drew me in and made me pay attention—now to his conflicted, yet deep, attachment to place.

Marguerite—intense and engaged—usually sat with us or stood at the kitchen counter, a cigarette or a paring knife in her hand, as she sliced vegetables, adding from time to time her own observations. They had met at the University of Michigan and were married in 1948. Marguerite talked about the way Brookside had become a home place for her as well as their children, about the Boston stage career of Jim's mother and her return to the South, about not growing tobacco ("It wasn't a moral thing, it just ruined the soil").

As graduate students years ago at Indiana University, we marveled at the two of them—an academic, glamour couple—and their five children. I could still see in Marg's attractive, aging face and white hair, the stunning blond beauty who dazzled the male graduate students at Indiana. After every class Jim taught, we asked one another how well he had performed. He was our academic celebrity, and she was his striking match, who we all wished would visit the English Department more often.

"I knew I would never sell the farm, but I always knew I would not stay here. But somehow I also knew, if I lived, I would come back." As a young man, he felt more of "a negative relation" to his home place than "a positive attraction," and he "never yearned to come back." It was just home. He did not speak about inevitability or destiny—as in landscape is destiny—but I heard something of that; as if, given the past and place he was born to and being sole heir, he could not really choose. "It was the thing I was somehow going to have to deal with." Instead of sentimentality or nostalgia, I heard Jim describing a conflicted attachment to the farm, quite consistent with his intellectual vision.

Jim's family history goes back to a Revolutionary War lieutenant, who received a grant of land along the New River and settled in Grayson County near Independence, still the closest town to the home place and about ninety miles from where I was living in Giles County, Virginia. Jim's strongest sense of family, however, began with his grandfather. Melville Beveridge Cox enlisted in the Confederate Army in 1861. After he was shot through both legs in 1863 at the Battle of White Sulphur, he returned home to Grayson County where he began acquiring land in the waters of Saddle

Creek that would become the 447-acre Cox farm. He would construct a house, a mill, and a number of out buildings and operate it as a family farm.

Built around 1877, Brookside Farm is Jim's birthplace and family home—*his primal place.* "I remember every square foot of this place. I've been here from the beginning," he explained. "The house was always a full of people, always. My mother had to cook and garden for the entire tribe that just poured in here. They didn't stay a day or two. It was always a week."

The farm was a typical nineteenth-century, southwest Virginia farm that supplied most of the family's needs. They raised beef cattle to sell for cash in a largely cashless economy. They kept cows for milk, as well as chickens, pigs, and sheep. They grew most of their own food. "We ate ham, pork, and chicken through the winter and chicken in the summer. We also had sausage in the winter because we had no refrigeration and no electricity until 1937 when I was twelve," Jim said.

"I worked so hard as a kid all summer in the hayfields, cutting, gathering, and stacking the hay, shocking some to put up in the barn," he continued. "We used teams of horses for all our farm work. Driving horses is a big thing when you're a kid. You learn to handle a team pretty early—how to go out and get them in the pasture, bring them in, harness them, and then hitch them up. We never did have a tractor or hay baler."

Jim explained how the farm worked when we walked from building to building and through nearby fields. We would look into the red-brick, pitched-roof spring and smoke houses; pass the old frame, shed-roof henhouse and the saddle-notched, drive-through corncrib; and then head to the red, gambrel-roofed barn with its open hay mow door and on to the old mill. Jim identified with these buildings. They represented in material terms the home, farm, and rural world that shaped the boy and man he became. He wanted me to see the structures and imagine the life he'd led and know the place he'd returned to.

The mill is an "astonishing" structure, according to the National Historic Register. It "was probably finished before the house," Jim explained, "and it began operating right away in 1877." It was painted a deep red on the outside—well kept in that sense—but the interior and the operating machinery had really impressed the Register: "Nearly all of the equipment survives in situ, and in good condition." Even the French silk sifters were still intact.

Inside, the mill was dry, dusty, dim, and so quiet with a faint smell of aged wood in the air. The only sound was Jim's voice as he showed me the equipment: the two pairs of millstones—one for wheat, one for corn—the roller equipment and its system of belts and belt wheels, the bins, hoppers, troughs, storage units, and so on. Then outside, we walked past the mill wheel and the millrace coming from Saddle Creek. "With a little repair and replacement, this mill could probably operate today," he suggested.

We passed most of our time together in the house where he was born and grew up—his primal place—his original corner of the world. If I may be philosophic for a moment about houses: "Our house," Gaston Bachelard writes, "is our first universe. . . . It is body and soul. . . . [It is] the topography of our intimate being." In Christian Norberg-Schulz's terms, "The place is the concrete manifestation of man's dwelling [in-the-world], and his identity depends on him belonging to places."

The style and structure of the house were typical of late nineteenth-century farm dwellings in southwest Virginia. It is a three-story frame structure built into a bank so that the front, facing the road, shows just two stories. It is three rooms (or bays) wide. On the second, or main, level, the central bay is the entrance hallway that runs from front to back and opens into each room. A shallow extension to the rear contains smaller rooms that also open onto the hallway, or central passage. It is early Georgian in style—or a modified I-plan house. Tall chimneys stand at each end and serve as interior fireplaces. A two-story porch extends fully across the face of the house. Like others from the era, it fronted directly on

Image 2. Brookside Farm. Image by author.

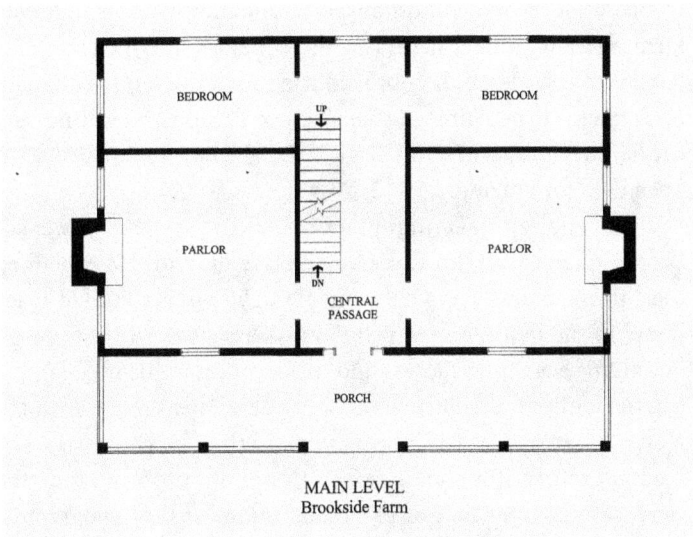

Ground plan 1. Brookside, main level. Chris Hudson, Shelter Alternatives.

a road—the Grayson County Turnpike. This common vernacular architecture characterized many of the houses in Giles County, where I lived and which I will write about in later chapters.

In 1918, the family opened an automotive service station adjacent to the house and along the turnpike. That board-and-batten frame building still stands, but it's not been a service station for many decades. The mill and the station served both the Cox family and their neighbors as part of a modest cash economy. In 1932–33, a paved U.S. Route 58, stretching from the Virginia coast to its coalfields, bypassed the house and station and, in doing so, anticipated basic changes in rural life.

These family farms and associated small businesses had largely disappeared by the mid-twentieth century. Farmers were earning money from off-the-farm work. More and more food became readily available in stores and supplied what people used to raise on their own. Food could be better preserved with electricity and refrigeration, as roads and transportation improved. Small family farms were no longer necessary or profitable. The mountainous terrain of the region did not lend itself to cash crop production. So farmers turned almost exclusively to raising and selling beef cattle. They grew corn for silage, grass for hay, and maintained open land for grazing.

Even though farming practices changed, the power of Brookside Farm as Jim Cox's home place did not. He explained that "the farm was very vital to my thoughts—it is a kind of space [here] in the mountains which I always associate with freedom." For all his academic success and his devotion to his own family, Jim must have thought of himself as a permanent outsider in the North—ready to bolt if it became more than he could stand or ready to return if for some reason things did not work out. The farm gave him a sense of psychological, as well as geographic, space and economic security. "You don't have to take it. You don't have to be abused because you have a place to go. If you got in a pinch, I could really make it here."

The home place also gave Jim a center—a heritage and an identity—yet it caused him great anxiety, made even more intense because of the way he became heir. "My brother was killed in a laboratory accident in Michigan. He was electrocuted in 1943. I had just finished my freshman year at the University of Michigan. I went into the Navy that fall. All that time, I knew I would probably come back. I knew I was going to inherit the farm if I didn't get killed myself." From that moment, Jim not only became sole heir, but he inherited the past and tradition of his family. It gave him security and independence. It also imposed a burden he would never be free of. It forced him into a relationship with his family and the South he might otherwise have eluded.

Jim knew the South well and thought often of his family and his southern identity. But he did not sentimentalize it. "A southerner should be ironic about his own country," Jim explained, both the North and South. A young man grows up southern but resists immersion in that identity, keeps his distance, and then leaves. He spends most of his adult life in the North, and in many ways shares northern liberal values, yet recognizes he is always an outsider— and a southerner. Eventually he returns. He comes home to his primal place.

His parents were liberal and committed to the New Deal. They rejected "this Dixiecrat crap," according to Jim, as well as the history of slavery, the Civil War, and segregation. "Any liberal in the South already feels morally guilty." So even before he went North, Jim felt somewhat estranged from the history and values of the South.

He also recognized the comic and ironic, as well as the painful, aspects of the southern boy, with his accent intact, going North and then having to prove himself. "Yankees can be just awful, being complacent in their beliefs. They hear a southern accent and you practically have to prove you have a mind!" He also had to prove his politics, since in those days, many northerners believed no one from the South could possibly be liberal. "I wanted to be liberal. I wanted to be northern in that sense. But a southerner sort of wants to prove he's more liberal. He has to outdo. He says, 'Goddammit, I'll show you how to be liberal.'" He assumed a "spirit of perversity," challenging northern ignorance about the South, making race much more prominent in his teaching than he had before, and sometimes, in silent amusement, simply letting northerners make fools of themselves.

"Anyone who has lost must have some humor. In a way, the southerner can't do anything but remember." Without humor and irony, however, the loser becomes obsessed with the loss and keeps replaying the game, reenacting the war, as if to say in losing we achieved our finest hour—courage, loyalty, determination—and actually "won." The loser becomes sentimental and nostalgic and invents a history and a war that somehow makes him proud.

At some point, "you just have to throw it over and wade out of it." You "would know somehow that your people were on the wrong side." You can't defend slavery or segregation. You understand the cruelty and destructiveness of the war. At the same time, you understand something about blood relationships and kinship the northerner doesn't. You have a sense of *place* few northerners have.

You can live a life defined by irony—by desire and resistance, attraction and distance, sympathy and candor, understanding and realism. You can live both in and out of the game, watching and wondering at it, laughing and anguishing over it. You also live knowing you have a home, a place, and a past. You are secure. You are free—free *from* the past and place, yet free to live *with* the past and at your primal place, passionately engaged, valuing your home and the South greatly, but mindful and ironic about it as well.

During his academic career, first at Indiana and then Dartmouth, Jim and Marguerite didn't think much about actually living at the farm. "I was totally involved in teaching. I always knew it was here, but I never yearned to come back. It was, however, the thing I was going to have to deal with. That was one reason I didn't like to think about it because I knew it was going to be tough."

"Coming back has been somewhat discouraging because there is less cleared land. A lot has grown up. I was so critical of the farm when I would come back and see it. I could just see my father was falling behind. We modified and changed a few things here, but I can't tell if I'm doing any better with it." He and Marg tried to rehabilitate and maintain what was left, and he acquired sixty additional acres, but it was hard. "It'll kill me like that. And I certainly won't kill it." Ironically, Jim came back to land that resisted *him* as much as he'd resisted it over the years.

Jim was conflicted about the South and his home throughout his time away, and when we talked, he still seemed to be. He wanted to return. He also had to return. The farm seemed to demand it. "It's just difficult to explain, but knowing I'm the sole survivor and my parents are old. Well, I saw it coming. I knew I had to deal with it and I knew I wouldn't sell it." He was contending physically, emotionally, and morally with a farm, his Cox family, the South, and the past, yet he loved the place and people he had come from. Brookside Farm had become his destiny.

Jim's experience of place was complex. Brookside was both historical and contemporary. It was land and buildings and the action and events that took place and were taking place every day he and Marg lived there. It was also psychological, social, and cultural. It was dynamic in the sense that it had changed and still was changing. Yet throughout time and the various changes, it remained his home place. It was simultaneously both different and the same. The farm has not only determined his life as a youth and young man; it had also deeply influenced his sense of self and shaped him socially and culturally.

———

My own sense of place is complex as well. Place refers to geographical and constructed places—location, topography, landscape, and buildings. It also refers to the psychological, social, and cultural influences at work at a given location. Place is ongoing and dynamic. It is an event as well as a location. These elements act in concert as forces rather than as separate, static features. Through their action, they constitute a place. The primary places we inhabit shape us profoundly. Seamus Heaney captures the interaction perfectly: "I was there / Me in place and the place in me."

Places also tell stories. They express certain forms of life and sustain those forms through the places people inhabit and the social and cultural structures they live within. Places enact the forms by structuring how people live and by influencing each inhabitant.

At Indiana University, Jim Cox had helped me learn how to teach and think. He taught me to read closely and intensely and to teach passionately and boldly. As a graduate student, I grew intellectually and professionally through Jim but not in a way related to place.

In Virginia, he helped me learn to think in new ways about the past and place. I began to understand his deep and conflicted attachment to place, his sense of the past, the ways Brookside Farm had shaped him, and why he had left and then returned. He forced me, in effect, to begin thinking about my own past and the places in my life. Jim's story touches every experience and concept in this book. My story—or better, my understanding of it—begins with James M. Cox and the mountain valley where I had recently come to live.

2.

Placeless in America

When my wife, Barbara, and I moved to Clover Hollow in 1991, we knew nothing about the history and people there, nothing about the nearby village of Newport, Virginia. We were simply searching for a sanctuary—a place quite separate from my demanding job at the university. In our quest for peace and beauty, we were perhaps doing nothing more than adding to the "PhD pollution" of the valley—as some natives saw it. Our situation differed dramatically from Jim Cox's at Brookside Farm.

We tried from the beginning, however, to learn about the place we had come to. We were trying to fit in. And I struggled with the problem of fitting in the entire time I lived there—do we? can we? if so, how? We built a house that paid homage to traditional farm architecture but was also designed with contemporary features. We built where nothing had been constructed before and transformed an open space into a place of human habitation. Over time, we not only learned about the valley and the village, but we also contributed modestly to its well-being by participating in community projects and writing about the people and their homes and histories. Eventually, we were accepted as valued "come-heres."

Clover Hollow and Newport, however, did far more than simply accept me. Along with Jim Cox, the people and the place showed me I'd been misreading my own life for decades—always thinking of myself as a man leaving his pasts and happily moving on. When I described myself that way to Jim, he looked at me, bemused and skeptical, and expressed a wry curiosity about my sense of self. It struck him as so American, or better, so northern and middle-class and so unlike his own experience. Saying nothing more, he left me to discover for myself the power of past and place.

I'd been writing at the time about always leaving my pasts and moving on—at one with Talmage Stanley's sense that "American middle-class culture . . . conveys messages and lessons of moving on. . . . These lessons of placelessness go to the root of who we are as Americans." I had gladly left my hometown for what I thought was a bigger and brighter life elsewhere. I had lived in ten different locations in five states before I moved to Clover Hollow.

My family was rooted in places, I assumed, much like everyone's. But I did not feel any connection to ancestral lands or memories. I would hold documents from my family's past and try to imagine myself part of those histories. I would scan the "Carlisle Genealogy" and feel little connection. James Carlisle emigrated from Loudoun County, Virginia, to Chillicothe, Ohio, in 1800. Five years later, he moved to Highland County, and before his death, he'd acquired two thousand acres. The deed books in Hillsboro, Ohio, show that the last parcel passed from the family in 1943.

Then, I would turn to the longer "Lucas-Bridges History" about my mother's family. It begins in England, passes through Canada—first into the Maritimes, then to Ontario—and ends in the United States with the immigration of my mother's parents. I read details about my grandfather's boyhood and youth. I found my mother on page 8 and references elsewhere to me and to my children. But I could not identify with any of it. I felt no trace of that history in me.

These documents, piles of photographs, and my own memories did not yield a sense of family or tradition that sustained me and enriched my imagination. I possessed memories and documents, but I did not have a heritage—or so I believed. I was missing something—something deeply important to my Clover Hollow neighbors and their ancestors and to Jim Cox: a sense of place and of the past in place. I had, in fact, been living a placeless life for decades.

The ten years I lived in Clover Hollow and near Newport gave me the most important community experience I'd known since my hometown. The people and the places of the valley and village enabled me to see my own past and primary places in a new light. Their ancestors had migrated into the area in the eighteenth and nineteenth centuries. The place profoundly affected their way of life and shaped their very identities, as the place had for their descendants and the family members I knew. This is what I wanted to understand—what it means and what it's like to be so deeply embedded in a place and its past.

Clover Hollow was the beginning of a long journey. I traveled the road back with several people—Jim Cox, my Clover Hollow friends, and my father, Ervin Carlisle, who died at age forty-six while I was still a teenager and whom I never really knew as an adult. Jim's story helped me understand the power of place and the imperative to understand one's past. It's as if he took me by the arm, pointed the way, and said, "Go." My Clover Hollow neighbors were living what I was seeking for myself—a place, a home, a true dwelling. My search for Ervin Carlisle, as a man and not simply my father, restored him to me in ways I could not have imagined. It also gave me back my hometown, Delaware, Ohio—then, a small, midwestern college town of 9,500 residents. That town remains my center and point of reference—my primary, even my primal place. I see—I now understand—with the eyes and imagination of a person largely made by Delaware.

———

I am writing to discover, define, and disclose. I am trying to make my experience intelligible to myself and to you, my reader. My desire and need for an identifiable place—a primal place—and for a sense of self, achieved through place, might speak to your own desire and need. I am making a Whitmanesque turn—"What is it then between us?" Place binds us together. The places and experiences of my life will likely differ from yours, but, if I am fortunate, mine might resonate with you or at least turn you back to your own experiences of place.

Hollow

3.

Clover Hollow

Our Sanctuary

It was a cold winter day—the temperature well below freezing. We were driving into the Allegheny Mountains to look at an eighty-acre parcel in Clover Hollow, a small valley near Newport, Virginia. Four of us—Sandy, the realtor; Barbara; me; and my daughter Janey—were so squeezed together on the bench seat of an old, four-wheel-drive pickup truck that Janey had to shift positions each time Sandy shifted gears. The main roads were clear, but as soon as we turned onto county road 601, we were traveling on packed snow and ice. At the foot of the Hollow, next to the Farrier farm, Sandy stopped, climbed out, switched the front hubs to four-wheel, climbed back in, and then crept along the glazed road for the next five miles. It was not promising. Another property we had tried to buy, five minutes closer to town, seemed to stretch the limit. Now, we were crawling along the ice on a forty-five-minute trip to the far end of Clover Hollow. Barbara was anxious. Janey seemed puzzled. I was wondering why we were doing this.

Without more than a slip or two, the truck climbed the crude road the owner had bulldozed to a meadow two hundred feet above the paved county road. The rush of wonder and excitement I felt seeing the snow-covered valley and mountains in the sunlight swept away all my doubts (well, most anyway) and virtually made the decision. Within weeks, we had acquired eighty acres of mountain and meadow. We were, it is true, fifteen miles and twenty-five minutes from work. Late at night, it seemed even farther. Our access road was sometimes impassable in winter. The land maintenance requirements sometimes overwhelmed me. Yet it was the right place.

Years before, Barbara had told me about the aristocracy of effort. We were canoeing across a Killarney Provincial Park lake in Ontario on a cool, bright August day toward a small island in the middle. After three difficult portages, we could neither see nor hear anyone else. We were exhilarated by the beauty of the lake, the steep granite cliffs falling into it, and the wooded island ahead, which on that day would be ours alone. We were aristocrats in a canoe.

Each December, I would recall that first moment, seeing the mountains and valley from the truck cab and then walking to the edge of the meadow to gaze in wonder. I remember the silence— interrupted only by our voices and crunching steps in the snow. We'd found our sanctuary. During the ten years I lived there, I saw a different landscape every day, as light, weather, and seasons changed, and I heard new sounds or heard sounds differently each time I listened. But on that first day, I did not know what I was seeing—other than a stunningly beautiful landscape.

I didn't know anyone who lived in the Hollow or in the nearby village of Newport; I knew nothing about the history of the meadow where I stood or about the Hollow below; I didn't even know the name of the place. We acquired land as strangers with neither family nor history to connect us to the mountains. We were making a superficial, aesthetic choice. We were seeking the picturesque. Nevertheless, that first time, besides being just that, marked a turning point. The meadow was my purchase—my place for exploring the spaces of my life and for discovering a people, a history, and a place.

"Did you grow up on a farm?" Ed Givens asked me—wondering why we chose to live "way up there." We were, it seemed, halfway to West Virginia and, for most university types, in a world apart. "Well, we wanted to live in the country," I replied. "You sure are out in the country," he said. "They live," Ed later told a friend, "about as far up in Clover Hollow as you can get." And Ed knew exactly how far. A retired Southern States executive and a former member of the Virginia Tech Board of Visitors, Ed was born and grew up on a family farm in Clover Hollow. He lived in the old country way of hard manual labor until he went to college and on to his career. Throughout his life, he always thought, nevertheless, of Clover Hollow as home, his primary place in the world.

I did not grow up on a farm. I did pass countless hours in the woods and fields behind my house in Delaware, Ohio. David Smith and I roamed along the Delaware Run from Blue Limestone Park, the old quarry, into the countryside. I remember days and weekends at my grandfather's farm. He lived in the city and rented the farm to a tenant. I walked the fields, watched farm animals, played in the creek, rode a horse once or twice, and even drove a tractor. I read Wordsworth in college and graduate school. During my year at Ohio State, enamored by poetic visions of walks in the country, my first wife, JoAnn, and I rented a farmhouse halfway between Delaware and Columbus. We passed many satisfying hours walking along creeks, through woods, and across fields. I was captivated by the romance of the rural.

Barbara and I moved to the mountains as a romantic escape. We knew nothing about living in the country; we did not farm or keep animals; we did not know how. We put out rhododendrons and chrysanthemums, knowing deer eat them, and they did. We left the old, yellow John Deere we'd bought from George Atkins out in the rain and snow. My neighbor Vince laughed at my tractor driving. "He doesn't drive it fast enough to have a wreck," he once told someone. Once, when I was driving through the woods above the house, the tractor stopped abruptly, but the rear wheels kept turning. I got off, looked, and realized I'd centered the differential perfectly on top of a tree stump. When George came to help, he smiled, but he must have laughed all the way back to his tractor dealership.

A few years later, when I was shopping for a newer tractor with a front loader, Marty Farrier, a neighbor in the Hollow, warned me against it. He was worried that I might kill myself driving with a heavy front loader. And I almost did—sliding down a steep embankment with the weight of the front loader pulling and the bush hog pushing. The slide stopped at the access road, but it took some time for my heart rate to slow.

I met Marty Farrier first at a university lunch, shortly after we'd acquired the land. At the time, he was on the Giles County Planning Commission and also the mail carrier for Clover Hollow. His Centennial Farm is situated at the foot of the Hollow. The road there passes through a narrow gap, alongside Clover Branch, and then opens to a dramatic view of Marty's 1905 house, faced by a two-story porch with large pillars. "It's fourteen or fifteen rooms depending on how you count," according to Marty. He and I talked only a few minutes, but he did tell me about the good people who lived up there. "They're pretty independent and usually stick to themselves, but when you need them, they'll be there."

Late one afternoon, I drove to the Hollow to meet Jay Williams, the contractor who was cutting a better access road from Route 601 up to our building site two hundred feet above the main road. A man was sitting in a pick-up truck at the foot of the access road. It was not Jay. That was my first meeting with Graydon Shrader. He'd come to explain that Jay was going to be late for our appointment. I don't know how long Graydon waited, but that was my first indication of how helpful he and others would be—even if I sometimes amused them.

Graydon farmed full time on land a quarter mile down the road. He didn't have time, it seemed to me, to wait around on some Blacksburg academic. But he did. After that, he and I became friendly neighbors. He offered helpful advice and assistance—"Make that outbuilding twice the size as you think it should be; you'll sure fill it up." Twice each season, Graydon cut hay in our lower and upper fields—"meadows" we called them. That helped us keep the fields clean and provided hay for his cattle.

Jay's concern, Graydon's generosity, and Marty's initial friendliness suggested that our Clover Hollow neighbors were welcoming us. Nancy Kate Givens's invitation to a July Fourth potluck a few months after we'd moved in confirmed our sense that we were being generously received. She lived about a mile away on the backside of the Hollow. We met neighbors there who would become friends and whose lives and histories differed dramatically from mine—among them Nancy Kate, Bill and Caroline Vincel, Mark and Melodie Givens, and Doak and Alva Lucas. I learned then that Nancy Kate had been looking into the history of the Hollow and its people for some years. As my interest in learning about where I was living developed, she helped me with my own writing about the Hollow and Newport.

Barbara and I attended our first Newport Village Council meeting that same July. The council had appointed working task forces to study and improve the village and its surrounding locations, including Clover Hollow and other nearby areas like Spruce Run, Sinking Creek, and Plow Screw. Groups were working on cleaning up and bringing business into the village center, restoring the 1916 covered bridge over Sinking Creek, developing the community center beyond Friday night potlucks and bingo, improving highway safety, increasing school support, studying Newport history, and so on.

Every type of resident attended—one hundred people in all—including Darryl Martin, the council organizer; Doak Lucas, a descendent of an early Clover Hollow family; Marty Farrier; university faculty who had lived there for years; a graduate student interested in rural communities; and recent newcomers, or come-heres, like Barbara and me. I heard accents from the North, the South, and the mountains. I saw people in every kind of dress. All shared a common purpose—a determination to restore a strong sense of community—and all believed they lived in a special place.

————

I realized in these early months that there was far more to Clover Hollow and Newport than the valley's beautiful landscape and the village's quaintness, however appealing or distracting both might be. I not only wanted to learn more; I also felt an obligation to understand the place and its people insofar as I could—people who'd been welcoming and generous, some fifth- and sixth-generation residents—a place with its own distinct history and identity. I did not want simply to appropriate and use the land for my purposes alone. We were trying to find a place—*make* a place, really—in a land where we had no claim to anything but property.

4.

Three Meadow Mountain

Homage and Innovation

When I drove to my property early on, I was completely unaware of the rich history and vital present I was passing by. It was all very picturesque. "Decorator cows," we'd say about the cattle standing on a hilltop silhouetted against the setting sun. Once I learned more about my neighbors and their family histories, however, I began passing *through* history. It captured and embraced me. The Hollow became deep and thick with memory and meaning.

I would turn off Route 460, drive through the village of Newport—now alive for me with stories about its incorporation in 1872 with 1,532 people; stories about a village with hotels, a grocery, hardware, feed, and drug stores; with mills, a tannery, and a blacksmith; and with tales about the destructive fire of 1905. I would then turn left onto Clover Hollow Road and head for home. I would pass the old milldam on Sinking Creek (it had had a second life providing electricity to parts of Newport) and then drive by the iconic covered bridge—the Red Bridge. Ed Givens remembered watching its 1916 dedication. Then I'd follow Clover Branch through a narrow gap where I occasionally saw Virginia Tech students studying the uplift in certain rock formations.

The gap opened into the Hollow itself and onto Marty Farrier's grand house. Just past Marty's, the road divided and formed a long loop around the Hollow. Holding to the right, I would pass the 1916 Floyd Harvey Givens house—Mark Givens's grandfather, then the "Fairbanks Standard [Livestock] Scales" owned by the

Doak Lucas family. "We used to weigh hundreds of heads there each fall," Doak told me. His 1861 farmhouse stood across the road at the end of a long drive. A handsome watercolor of that farm hangs in my study. Doak is gone now. The farmhouse sold. His son, Jimmy, and his wife, Gail, have moved into Blacksburg where Jim, as he's known there, owns an appraisal business.

Farther along, I would pass the white frame Clover Hollow Christian Church, constructed in 1862 and restored in the 1920s but no longer used as a church. Then I'd pass the road to the Clover Hollow Cemetery—all those names and generations. Barbara Carlisle now rests there. Then came Jake Sibold's large, handsome frame house built for himself and his wife, Eveline Surface. They came to the Hollow in the 1840s and lived first in a log house. Their home stayed in the family until Nancy Sibold's death, generations later, when it was sold to a newcomer family. After two ninety-degree turns, I'd drive by the former Upper Clover Hollow School—used as a hay barn for decades but altogether gone now. Then one more turn, at the Crockett Sarver house, and I would head on toward Three Meadow Mountain. I'd see my 1905 barn first on the right—still somewhat usable then—and across the road the original farmhouse, built by Harv Link in 1905. Just past the barn I'd turn into my drive and ascend the winding road for almost a half mile to my house.

On that daily drive, I traveled through the past and present of life in Clover Hollow—a life of loss and living, of change and continuity, of people and places who trace their ancestry back five, six, and seven generations, of newcomers who were helping shape the present and future of the Hollow and who were beginning their own stories. I also felt my own dilemma. No longer as ignorant as I was, I was still a come-here, trying to make a place and fit in.

From the moment we acquired property in Clover Hollow, I'd been anxious about *how* we might fit in. I thought about this, first, in architectural terms aware of the vernacular farmhouses I passed every day. A building, Witold Rybczynski suggests, which "ignores its context [and which therefore does not fit] . . . lacks a crucial ingredient—*meaning*" (emphasis mine). But fitting in is more than architecture. It is a complex personal, social, and cultural matter as well. It is also like that with people. Unless we learn the context—the language and stories of the place we live— our own stories will be insubstantial and incomplete.

At some point in October 1992, I noticed a white spot on a distant mountainside whenever I stopped at Super-Valu—the grocery and gas store at US 460 and Virginia 42. I paid no attention at first, but then I realized it was my house under construction. The white, Tyvek house-wrap made it stand out, even from five miles away. Then in late December, the house receded from view—the mountain seemed to absorb it—after the builders covered it with gray siding. It was visible long enough, however, to make me wonder if this house—and if *we*—might fit into the landscape and the community. The architect had said the meadow would never be the same after construction; the excavation proved him right. I also knew that a house placed "up there"—high on the mountainside—would change the head of the Hollow significantly. It was visible along the road from a mile away, and at night the outside lights could be seen from three miles. You could not miss it.

The house did announce our presence, but in the end, we did not alter the view *so* dramatically. Those few months of a white house, however, forced me to consider the entire Hollow in contemplating how we might fit. Thoughtful regard for the design and the site were no longer sufficient, nor was it enough to think only about our location in the head of the Hollow. We were not just building a house on our own private property, isolated from everything and everyone else. We were entering and, in some sense, modifying a well-defined domain. I felt an obligation, therefore, to consider not only the buildings from one end of the valley to the other, but also the history, the way of life, and the people of Clover

Hollow. And that meant I had to think about us as well—how we might alter the social landscape and the natural terrain. In short, I had to consider the *place*—Clover Hollow—in all its complexity. That would take years.

The style of Three Meadow Mountain—the name we'd given the house and land—is based on the traditional western Virginia, T-plan or L-plan farmhouse of the nineteenth century. We chose the vernacular mainly for intellectual and aesthetic reasons—so typically academic—but we also believed the house might fit better into the landscape and community if we paid homage to the history and culture of the place.

The earliest houses built by European descendants in western Virginia were often one-room log houses with an end chimney for the interior fireplace, usually the only source of heat in cold weather. Owners often expanded these houses by adding a second room and a second floor constructed still of logs. Some were later expanded into a full I-plan house—one room deep, three rooms (or bays) wide, and two stories high. Kitchens were sometimes placed in separate, adjacent structures or later appended at the rear of the main house. Owners often constructed the third bay in a balloon or frame style. They then would cover the entire structure with clapboard siding to unify the whole.

Later in the century, the classic I-plan house further evolved into a T-plan or L-plan structure. T-plan houses consisted of three bays across (the standard I-plan form) and two bays extended to the rear. The middle bay—the front hall, an open two-story space with a stairway—was common to both the front and the extension. These later forms were often constructed all at once in a frame, or balloon, style. They were typically fronted by either a two-level porch and portico extending forward from the middle bay—as in the 1892 Givens house, which I describe later—or by a two-level porch extending across the entire front of the house—as in the Cox house at Brookside Farm. Some architectural historians

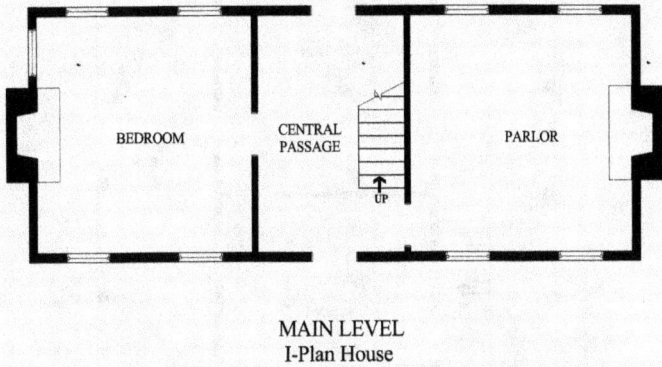

MAIN LEVEL
I-Plan House

Ground plan 2. I-house ground plan. Chris Hudson, Shelter Alternatives.

---

might describe it, alternatively, as early Georgian. L-plan houses simply extended the two rearward bays from one of the side front bays.

The inside of both I-plan and T-plan houses was closed. "Upon entering," Henry Glassie has explained, "you do not stand in a room where people sit. You are in an unheated, unlit corridor—a hallway, a way to the hall—out of which you must be led to the sociable place." That is, visitors entered into the hallway (the passage) and could not see into any rooms unless the family opened the doors into the two side rooms or into the dining room. The structure separated inside from outside and effectively controlled entry or protected the family. Family and friends would typically enter through a side door into the kitchen.

Three Meadow Mountain adapted the L-plan, with three bays across the front, two bays extending to the rear, and a two-story porch and portico. It broke sharply with tradition, however, by

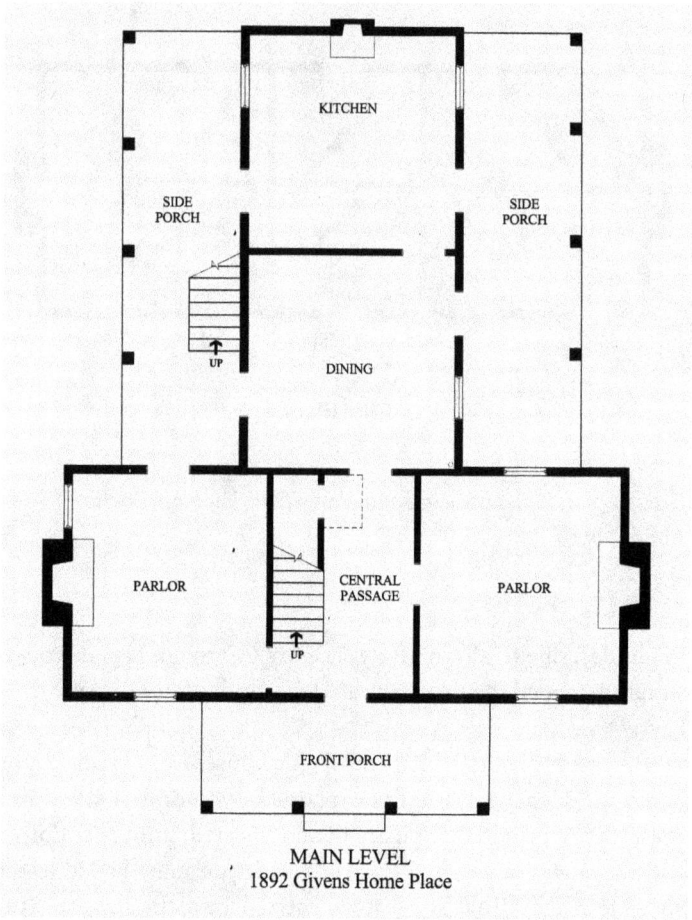

KITCHEN

SIDE
PORCH

SIDE
PORCH

UP

DINING

PARLOR

CENTRAL
PASSAGE

PARLOR

UP

FRONT PORCH

MAIN LEVEL
1892 Givens Home Place

Ground plan 3. T-plan house, 1892 Givens home. Chris Hudson, Shelter Alternatives.

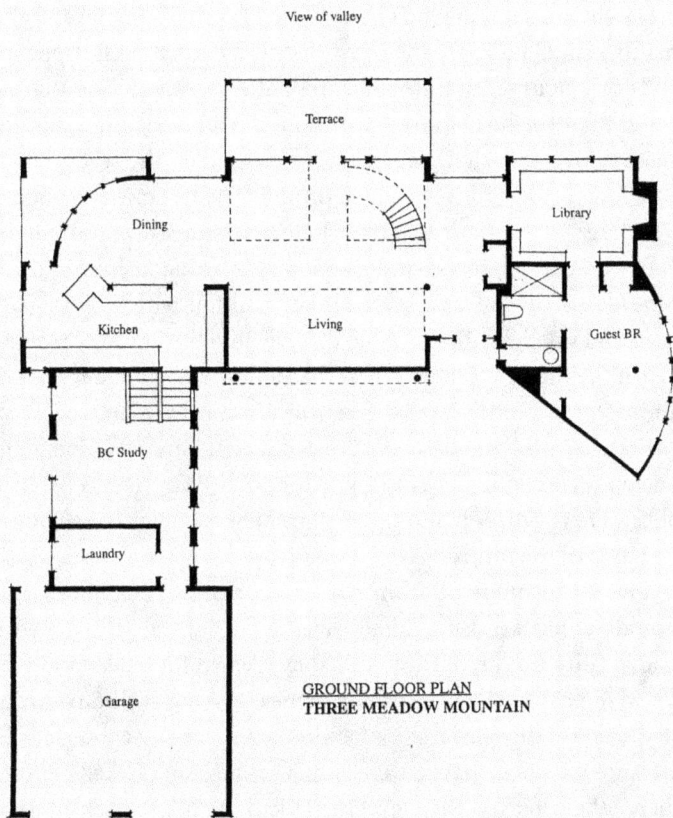

View of valley

Terrace

Dining

Library

Kitchen

Living

Guest BR

BC Study

Laundry

GROUND FLOOR PLAN
THREE MEADOW MOUNTAIN

Garage

Ground plan 4. Three Meadow Mountain, first floor. Hayden May, architect.

incorporating a number of contemporary variations. The architect broke the conventional box by cutting off corners on one end and making small porches, and by extending a corner on the opposite end with a curved wall that swung out, so to speak, for a larger room than the box would have allowed. He opened the interior space with a two-story central room crossed on the second floor by a long bridge between the bedrooms. He designed the house with huge windows—two stories high on the face overlooking the valley and large second-story windows opening to John's Creek Mountain. The entrance also broke tradition with a clear glass door and sidelights that opened directly into the house.

Unlike the newcomers across the Hollow who built a replica of a nineteenth century farmhouse (with ironically, a massive Wolf gas range in the kitchen), we created a contemporary home but one with clear references to the old houses. We tried to balance old and new—to look into the past, rather nostalgically I suppose, and toward the future.

From the moment framing was complete, neighbors and friends could see that Three Meadow Mountain resembled a classic Giles County farmhouse. They seemed to understand that we were paying respect to their homes. During the house-wrap stage, especially, passersby often looked twice—new house? old farmhouse?—before they were sure. Our neighbors liked coming there. They enjoyed the view. They liked the house itself in both its vernacular and contemporary modes. On first impression, anyway, the combination seemed agreeable.

I would *like* to think that Nancy Sibold, then a fifty-year resident of the Hollow who had married into an old farming family, spoke for many: "I think it's good in a lot of ways. I like to meet new people, like when I visited you in your home. You were all very gracious, and I really enjoyed that. . . . It hasn't bothered me at all that people have moved here. I think everyone has welcomed them." Mark Givens, a young farmer I came to know fairly well,

Image 3. Three Meadow Mountain facade. Image by author.

also assured me that we have a place—sort of. "I don't see y'all as invaders. Y'all haven't tried to change anything or bothered anyone." Mark was trying to be friendly, but I did hear an implied sense of difference in what he said. We might have been *in* the place. We might have been acceptable. However, we were definitely not *of* it.

There is, however, much more to fitting in than first impressions or polite acceptance. Terms like "propriety" and "decorum" help explain architectural fit. The designers of a house must learn the architectural language of the houses and buildings around it to achieve significance. Three Meadow Mountain attempted this, but it did not simply repeat the past. The old houses were direct adaptations to a rural, mountain situation. People used materials from the land—stones for underpinnings, clay for bricks, logs for walls, and, later, lumber milled on site for framing and siding. They built and expanded the houses as need required and money allowed—starting, in the early years, with a log structure, then adding a framed addition and covering it all with clapboard. The houses were small and required families to use most rooms for several purposes. They suited a farm family's needs well, and they also had a pleasing aesthetic. But Three Meadow Mountain was not a nineteenth-century farmhouse; nor were we farmers. The house emphasized beauty and historical reference over simple practicality. Nevertheless, we tried to learn the language and speak it in a new way.

We were likewise trying to learn the social and historical languages of Clover Hollow and Newport so that we could understand the community and find our place in it. Propriety and decorum are necessary for this as well. If we wanted to be accepted and in some way included—and not regarded simply as intrusive outsiders or "flatland furriners," as Doris Link, a sixth-generation native, laughingly referred to newcomers—we had to find appropriate and respectful ways to relate to people and their primal and primary places. Even those, like Doris, who are receptive to newcomers "don't," as she put it, "want somebody comin' in here and tellin' them [the natives] what's good for them when they pretty much know what's good for them." Out there, we were the ones who spoke with an accent—a social and cultural accent, as well as a linguistic one.

Architecture reflects our desire to shape the physical world to fit us—according to architect and author Paul Shepheard. It expresses our need to take possession of the landscape. Left only at that, however, architecture becomes a kind of colonizing gesture, characteristic of many newcomers in Clover Hollow. They come for the outdoors, space for horses, a few acres to keep a handful of livestock, room for gardens, or simply for the beauty of the mountains. They do not come to farm; nor do many have a family or cultural history to draw them there. These professionals and academics come to possess the land and shape the world to fit their interests.

Barbara and I were not so different from more recent newcomers. We acquired eighty acres of mountainside. We built an access road where none had been. We constructed a house where none was before. We changed the end of the valley noticeably. These were certainly acts of possession and even of displacement. But that wasn't the whole story.

People also have a compelling responsibility to serve and not just possess. Architectural space, "Existential space [architecture]," according to Christian Norberg-Schulz, "cannot be understood in terms of man's *needs* alone, but only as a result of his *interaction* with an environment, which he has to understand and accept. Architecture should serve the *public* world." It is the same for people. Barbara and I had a responsibility to Clover Hollow and Newport. We had an obligation "to protect and articulate the place [we were] given to take care of." Mark Givens once said it very directly: "You should want to preserve what you came here for." We took possession. We served. We then became participants in the well-being of the landscape and the community. We fit in. And in the end, we told a new story. That was, at any rate, my ideal—elusive and unrealistic as it might have been.

Three Meadow Mountain served by paying homage to tradition and community, and it also served by redefining the head of the Hollow in appropriate ways. The house, for example, centered and focused the landscape around it, much as Marty's red barn in the valley below provided a point of reference and structured our view of the Hollow. From a distance, for better or worse, the house gave people a new way of looking at the mountainside. The contrast between house and nature made each appear more sharply. The house also linked the three meadows—the field below, the meadow we lived on, and the large open meadow above us—and drew the nearby woods and the forests on the mountain toward it.

It organized the landscape and gave it fuller human definition. People had been doing that for centuries in this Hollow—making human places, orienting themselves, with paths, roads, homes, barns, fences, and fields. They were creating a social and cultural domain out of raw mountain space. They possessed, but they also assumed, responsibility for the place they had been given to care for. Our house continued a tradition, I would like to believe, by giving new focus, definition, and coherence to the head of the Hollow. Consistent with other dwellings, a house, if it is well designed and well placed, can serve and also improve its landscape. Its inhabitants, acting knowledgeably and mindfully, can as well.

As dwellers in the house and on the land, Barbara and I also interacted with the landscape. The architecture created a visual flow between the inside and outside through large windows that gave broad views in every direction and through exterior glass doors that provided visual, as well as physical, entrances and exits. The design extended the life within outward and brought the landscape inward. From inside, we saw fields, mountains, and sky—an endless variety of light and shadow. We felt warm sunshine through the glass. We saw and heard the wind. Open a window or step through a doorway, and we felt the wind and the warm sun directly. We sensed cold or mild air. We were refreshed by rain and chilled by snow. We smelled grass and trees. And we heard birds, cattle, tractors, trucks, and cars—the life of Clover Hollow. The design of the house dramatized the relationship between house, self, and place—between the human and natural worlds.

This kind of openness is hardly unique to Clover Hollow. It could be a feature of any house with a view. But there, for us, it provided the ground for our participation in the landscape and life of the Hollow. It was, so to speak, our first gesture of reaching out to the valley and its people and drawing the place and people in.

Marty Farrier came to the house one evening to help us make a list of invitees for an open house. We wanted to host as many neighbors and Hollow residents as we could—everyone from Marty's house at the foot of the Hollow to the head, where we lived—and even beyond us to the last house on the road up to Rocky Gap, where Laurel Springs Road passes over into Maggie, the next valley, and where the Appalachian Trail crosses.

"I'm not supposed to do this," Marty said, "so you mustn't tell." We were sitting at the harvest table in the dining area. Marty traced mentally his mail route—naming every person in every house who received mail. "Do you know them? They're real nice people. They'll come. . . . Now those folks probably won't, but you should invite them anyway. You should invite everyone. It's the gesture." With each comment we were looking at a different name and slowly working our way around the Hollow. "Her mother bought the abandoned house on the girls' property and is restoring it. She may still be living with them. I don't know whether she'll come or not. You might get two out of three from there."

And we continued along the route. "Well, you'd better not invite them. You know what they did?" We did not. "They" had persuaded the former owners of the property, where they were living, to carry a second mortgage and now refused to pay or move. At certain names, Marty paused and smiled, eyes bright. His look told us there was a story. It also said we would not hear it. The mailman knows everything; newcomers, almost nothing.

People did come. They came to see the house. They came to see their friends. They came to know us better. At a later time, Mark Givens and I were talking about his house—the family home place built by his grandfather in 1916. "I feel pretty attached to it . . . until I go in yours, and then I'd like to have a new one. I don't feel that way about many. I think you made yours livable. I really like coming to your house." That first gathering went so well we invited people every December thereafter.

One year, on the day of the open house, it had rained for thirty-six hours. Water was flowing across the junction of Rocky Sink Road and Clover Hollow Road near the foot of the Hollow. Zells

Mill Road, which parallels Sinking Creek, was closed opposite the Newport House. More water was crashing over the old milldam and tumbling down the creek than many recalled ever seeing. A number of people couldn't get through, or were they to try, they feared they might not get home. Water was rising everywhere.

One couple tried three different routes before they found their way over from Mountain Lake. Victor and Stewart Link arrived first. Their family had owned this property from 1905 to 1915—the Link Place situated at the head of Clover Hollow. They lived on Rocky Sink Road in Plow Screw near land an ancestor had settled in the 1790s. Others soon arrived. Doak and Alva Lucas and Caroline and Bill Vincel came dressed for a Sunday visit—the men in coats and ties, Alva in a suit, Caroline in a handsome dress and jacket. Graydon and Glenda Shrader brought a birdhouse that Graydon had made as a gift. Mark and Melodie Givens; their children, Daniel and Kelly; and Mark's mother, Mary Givens, came. The two children liked exploring the house—up the stairs, out on the upper porch, back down, and into the library and guest room.

Nancy Sibold wanted to come so badly, but high water over her road and bridge kept her home. Marty Farrier appeared in his pickup, not yet for the party but to chase five head of cattle that had broken through a fence and were roaming our land. "I'll be back," he said as he headed down toward the pond meadow. "After I do this, I have to get the womenfolk." A half hour later, I saw Marty and his son driving the cattle along Clover Hollow Road to another pasture. He never did get back for the open house.

People were making themselves comfortable—Caroline sitting in a huge chair and others gathering on the sofas. Mary Givens moved through the room talking with everyone she knew. Others collected on chairs under the open stairway—or stood near the kitchen at the far end of the buffet table telling stories about learning to fly, trips taken after retirement, a professor's archaeological adventures diving to shipwrecks off the coast of Israel.

Several mentioned how unusual it was for turnips to be good so late in the season. Everyone talked at one time or another about rain, mud, and high water.

The afternoon ended. The rain had stopped as well. People began leaving. They all made it home safely. I felt enriched by their presence. I sensed that we might find a proper place in Clover Hollow. We had extended our hands to our guests. Neighbors and friends had taken them into theirs and held them for a brief time. Our guests seemed pleased to have been invited. "You can tell when you are or are not being received," Ed Givens once told me.

Important as these encounters were, they were social and relatively superficial. I wanted to know more—both as a neighbor (but still a come-here) and as a student of where I was living. I wanted to talk with people about their lives and the Hollow. I was not bold enough to call or ask them face to face, so I sent letters. Most were quite receptive, and most of those let me record our conversations. This was simply the beginning. It would take years of conversations, research, and reflection before I understood the power of the Hollow as a place in the lives of the people whose families had lived there for generations.

5.

Clover Hollow

The Place

Clover Hollow is a small valley located northeast of Newport, Virginia. I visualize it as an oval lying southwest to northeast. It is two miles wide and three miles in length from the foot at Marty's house, to the head, where we lived. Big Ridge divides the valley and creates two long bottoms for farming. Johns Creek and Clover Hollow Mountains border it on the northwest and southeast. The two come together at the head where Kelly Knob dominates, the Appalachian Trail crosses, and Rocky Gap opens into the next valley.

Sinking Creek rises near New Castle and traces the history of people and places as it flows toward Newport. It runs alongside an early nineteenth-century road bed, passes the exposed underpinnings of inns and houses—long since gone—moves by collapsed barns and abandoned homes, streams past century-old dwellings, flows underneath a small covered bridge, and swings by the large stone furnace of an old iron smelt before it turns north, just east of Newport. There it spills over a concrete dam—built originally for a long-gone gristmill. A few hundred yards farther downstream, the creek passes under the 1916 covered bridge, which replaced a ford people had been using for a century. Sinking Creek bends again at the opening of the narrow pass into Clover Hollow and then courses on to the New River. To get into the Hollow, settlers left the creek, followed a small branch through a narrow gap, and then entered a beautiful, small valley—Clover Bottom, as it was first known for its rich, fertile land.

In the 1790s, the ancestors of the people I came to know were obtaining land patents nearby along Sinking Creek. These mostly Scotch-Irish and German families had migrated down the Valley of Virginia from the northeast. Although John Lafon probably acquired land in the 1790s, the earliest record I found shows a land transfer to him in 1811. Between 1811 and 1834, he acquired 990 acres at "the head of Clover Hollow" near the "head branch waters of the Clover Bottom branch," as the 1829 record of transfer describes it. The Lafon holdings included land at the head of the Hollow and up toward Rocky Gap, as well as acreage along Clover Hollow Mountain, opposite my property. It is not clear when a Lafon obtained the land just west of what was mine, but William Hale Lafon, the grandson of John, built a house there on a high knoll in 1853. The Lafons, who owned the land when I lived next to it, were seventh-generation landowners in Clover Hollow.

The Givens were early settlers of the Shenandoah—part of the Scotch-Irish migration down the valley. Daniel Givens, who lived near New Castle, died in 1822. His son, Isaiah, became a relatively well-to-do landowner along Sinking Creek. Isaiah's son, James Stafford Givens, the great-grandfather of the Givens I knew, settled still farther down the creek near Simmonsville, about eight miles from Newport. This line of Givens was moving generation by generation toward, and then with James Stafford's purchase in 1868, into Clover Hollow. He did not, however, live there himself. His three sons were the first of the family who did—James Monroe in 1873, Joseph Cale (Caroline Givens's grandfather) in 1878, and Floyd Harvey (Mark's grandfather) in 1916.

These early families were establishing footholds in a relatively harsh natural environment. They were transforming rocky clearings, woodlands, and mountainsides—land and space—into *places* with human meaning and value. The location and topography of Clover Hollow constrained how people lived—and still does. It is a terrain of limits. It was difficult to work and negotiate—especially in the early years. Even today, the land not far up the mountainside is uncultivated and wild—although more and more newcomers are building houses higher up. We built four hundred feet above the paved road at an elevation of 2,600 feet. The ridge above us stood at 3,400 feet.

The mountains dominated people. No matter how high up they built, the mountains rose even farther. But the valley was also a receptive space. The hollows and surrounding woodlands contained and sheltered. The slopes at the base of the mountains and the bottomland allowed for human habitation and cultivation. Clover Hollow is also beautiful and at times breathtakingly so. Unfriendly in its harshness and indifference, it was friendly in its fertility and beauty.

These families were also building houses—small, crude log structures at first, but they were nevertheless creating dwellings—distinctive places whereby they situated themselves, if simply and unknowingly, between earth and sky and created a world. Their homes provided modest footholds in a difficult environment. They provided comfort and a sense of order and security. They enabled a family to feel at home in the world. They centered the lives of the families on the land, in a house, and in a family.

The people were also establishing settlements to support and connect their independent and rather isolated farms. By means of paths, roads, and commerce, the settlements gathered many families together to form small communities, like Newport. The people were asserting a human and cultural presence of a particular character based initially on independent, family, and subsistence agriculture, conducted under difficult conditions, in mountainous terrain. Over time, they developed a powerful sense of common ancestry, shared experience, kinship, and rootedness that gave them a sense of place and a sense of identity. Their independent spirit, their egalitarian sense, and their capacity to join together in work and community helped them survive.

In my frame of reference, landscape was destiny. The place shaped a way of life for the settlers and their descendants, if it did not, in fact, force a way of life on them, much as Brookside Farm and Grayson County had for Jim Cox's family and their ancestors. They depended on the land not only for their subsistence but also for their very dwellings. They built homes from local materials: stone for foundations; logs and then milled wood for walls, framing, and siding; and clay for bricks. They expanded the houses as need required and resources allowed. They tapped their own springs for water. Sometimes, they employed skilled labor or bought materials locally, but they relied mainly on themselves and their land.

Small communities, like Newport, helped overcome the loneliness of modest, isolated farms and also enabled them to collaborate on the hardest tasks of farming. This way of life affected the people psychologically, socially, and culturally and led to the independent, self-sufficient, private, family-centered, and reticent character of the people and their way of life. The place *was* the thing. It enacted a particular form of rural life. Clover Hollow—Clover Bottom—provided sustenance, dwellings, and society. It also limited the possibilities sharply. The landscape—the place—disabled as well as enabled.

6.

The 1875 Lafon Home Place

Two abandoned houses were still standing on the Lafon land just west of mine in the 1990s. William Hale Lafon had built the smaller of the two in 1853. He fathered sixteen children, the oldest of whom, James Harvey, was born in 1853 when his mother Lucinda Jane was just fifteen. When James Harvey married in 1875, he constructed the second house in a hollow just below his father's home.

Roaming along the mountainside one day, soon after I had moved there, I discovered the houses. They fascinated me in the way ruins interest us all. But I also thought these two houses could tell stories not only about the Lafons but also about Clover Hollow. And that sense prompted me to call Wade Lafon. Wade and his sister, Annie, grew up in the 1875 house but had left the Hollow decades before. Wade had moved to Narrows in 1964 to be closer to his work at Celanese—a producer of cellulose acetate tow, used in filter media. Annie had left before that for nursing school. Although both lived their adult lives elsewhere, they nevertheless regarded that collapsing house and overgrown land as home. "I'm going home," Wade would tell people when he left his house in Pulaski for Clover Hollow. And Annie confirmed that. "Sometimes [when] someone asks, 'Where are you going?' I say, 'I'm going home.'"

With materials mainly from the mountainside, James Harvey— Wade and Annie's grandfather—erected a two-story log house with a stone chimney. Later, he added a framed section. Still later, he constructed a kitchen appendage to the rear. The original kitchen had been situated apart from the house in what became the

Image 4. Lafon ruin, ca. 2000. Image by author.

family's garden. Subsequently, James Harvey covered the exposed logs, the framed addition, and the kitchen with board siding that unified the exterior. He also built a front porch.

When it was finished, the two-story Lafon home place consisted of an entrance hall and two side rooms on the first floor—a parlor that doubled as the parents' and Annie's bedroom and a family or living room that opened into the kitchen; on the second floor were two large rooms—one a bedroom for Wade and his brother, Lewis, and the other, an unfinished room for storage—the log walls still exposed. Only the kitchen and the family room were heated. James had constructed a dwelling that provided, for the time, comfort and a sense of order and security. It centered the family's life—and, in a sense, it still did when I knew Wade and Annie.

As a skilled carpenter, James Harvey also gave the house a simple beauty that enhanced its domesticity. I could see signs of that, even after years of deterioration, in the two broad window gables on the exterior and in the mantelpiece, the chair rail lining

Image 5. Lafon house, ca. 1920s. Courtesy of Wade Lafon.

the walls, the paneling below and wallpaper above, and also in the table he made for the parlor.

A clearer indication of this modest aesthetic appears in a photograph from the 1920s. A teenage cousin, Elva Sheets, wearing a white dress, is standing at the top of the porch steps looking toward the photographer. There are scrollwork braces between the posts and the porch roof; the porch railings are cut balusters. The house appears to be painted white, but the exterior framing—around the windows, the face boards outlining the gables, the porch posts, the cross pieces of the railings—is painted a darker color. In the center of the porch gable, someone has painted a large star. This "fancy work" (Annie's term) gave the front definition and clarity—even a kind of elegance. It is summer. The trees and bushes are in full leaf. Vines are growing up the porch posts and across the porch ceiling. Pots of plants have been placed on the outside of the steps.

The evolution of James Harvey's house and its architecture clearly expressed and enabled this form of rural life in the mountains. First, it was a "two-thirds" log house (two side-by-side rooms), then with the addition of a framed, third bay, a full I-plan house, later, an I-plan house with a kitchen appendage to the rear, and finally, a handsome farmhouse, covered with siding. Constructed with local materials, built largely by its owner, expanded as necessary and affordable, heated minimally, with rooms used for multiple family purposes, the 1875 house was itself a subsistence building, just as the farm was a subsistence farm well into the mid-twentieth century.

The house originally stood on what was then the main road. The old roadbed and a low stone wall alongside it were still visible. The house now sits well back and above the present paved road. When Wade and Annie lived there, the family kept the land clear all the way to the new road. They walked down a path through the family apple orchard to get the mail or catch the school bus. "You could take a horse or a sled and go through here," Wade said. In my time, the path and orchard had disappeared in tangled thickets.

The Lafons owned seventy-nine acres on the side of Johns Creek Mountain where they practiced the diversified farming typical of the hollow. "It was basically a farm community," Wade explained, "and everyone farmed with horses"—until Mark Givens's granddaddy bought a Case tractor. The Lafons never did own one. The family kept cows and chickens for milk, eggs, and fried chicken. "We always had two good hogs," Annie said—which they slaughtered at Thanksgiving, when "Mother would make sausage and spare ribs." They raised a few calves to sell and grew corn, hay, and oats for livestock feed. Every spring they planted a large vegetable garden where they grew beans, carrots, corn, potatoes, lettuce, onions, tomatoes, beets, radishes, and cucumbers. They harvested cherries, peaches, plums, damsons, and apples from their fruit trees and picked blackberries and dewberries from bushes in the woods.

Annie and her mother canned fruit and vegetables just as her mother and Annie's grandmother had before. "You may not believe this, but when my mother died in 1981 and Lewis came here to clean up things, we had cans of food my mother and grandmother canned—and my grandmother died in 1936." During the winter, the Lafons relied on what they'd put up, stored, and cured. They also bought bulk supplies: "We would get in one hundred pounds of pinto beans, one hundred pounds of salt, one hundred pounds of sugar, and a barrel of flour," Annie said.

Their way of farming began to change after they acquired electricity in 1947 and bought a refrigerator. The cash their father, Wat Lafon, earned from periodic, off-the-farm work—on a tipple at a West Virginia coal mine and at the Radford Arsenal during World War II and the Korean War—enabled them to buy, and, with the fridge, store a lot of what they'd produced and grown before. They also began purchasing milk, eggs, chicken, and pork. At the end of their lives, their parents, Annie explained, "didn't have chickens, cows, or hogs. They bought all that."

When I walked with Wade and Annie around their home place, I saw an abandoned and decaying house, the remains of a stone wall and board fence along the old road bed, a pile of stones from the original chimney, untended irises, a stray narcissus, deep grass, tangles of weeds and briars, unpruned and broken trees, the outline of a porch long since torn down, bare weathered siding with only a faint residue of color on the front door (was it red?), rotting outbuildings. Thickets and woods in every direction. No open land anywhere. No farm. But through photographs and Wade and Annie's stories about growing up and living in the house, I could sense how rich the life in this house once was.

"We were never bored," Annie says. "We just had a good time. ... We had a good life growing up." She did work in the cornfield, she explained, "but for the most part, I worked inside. I liked the housework, so even when we were growing up and I was pretty little, I would stay home and get dinner. I learned to cook and sew. My mother taught me to knit and crochet." Wade, of course, helped with farm work. "I cut wood and got it in. We had some corn and oats. I milked the cows. We raised a few calves."

When they were young, they played like typical country kids. They hiked in the woods and would go hunting a bit. Baseball was a big thing for them. "We'd play ball down here in the yard. If we didn't have a factory made ball, we would make one of our own." The girls played like girls anywhere. "Out where Wilbur's house sits now," Annie described, "there was a huge hickory nut tree. The roots were coming up. We'd take playthings [out there] and make a living room and a bedroom. We'd take our dolls."

During bad weather or the cold days of winter, they read, "sometimes, practically all day," Annie said. "We used to get this little weekly magazine, *The Grit*. We bought books. We didn't have a radio early on, but we'd go up to our cousins' [the 1853 house] every Saturday night and listen to the Grand Ole Opry. We would have drinks and treats. That was really something."

Every summer for years, the Lafons held a family reunion in the yard of the home place. "One of the things I really used to look forward to, as far as a country boy would enjoy, was our family reunions down home because that was the home place," Wade explained. "All of this was cleared out," Annie was pointing to an area in front of the house. We were standing in tall grass and weeds, looking at an area a person could barely walk through. "On Saturday, all the relatives would come. Everyone brought a dish, so that the tables were loaded with chicken, ham, salads, sandwiches, cakes, pies, melons and lots of non-alcoholic drinks. We would probably have eighty to one hundred people. We had a real nice place for reunions." Eventually, the family moved their reunions to the Newport Community Center. One year Wade invited Barbara and me to join them.

Wade was drawn back to this place repeatedly even though, when I knew him, he had not lived there for thirty years. No one had lived in the house since 1982—and it was striking how rapidly it had decayed in only fifteen years. Nevertheless, it is their home place. Wade felt a little melancholy coming "home"—the land was so overgrown and the house badly deteriorated. "It's easier when I'm with my boys. When I come up here alone, I just try to keep busy." When he led me through the clutter on the floors, around old, worn furniture, under peeling wallpaper, he felt uncomfortable. "This is why I'm ashamed to bring you in here." Still, it was home.

Beginning with the 1790s, the Lafon family established a way of life that subsequent generations sustained—a form of life that provided their descendants with a home and a place that centered and guided their lives, and that still defined a vital culture for the living members of the family. Wade's adult sons had become owners of the house and land. "They think this is the next place to heaven," Wade said.

It's not like it used to be for the Lafons and for others who left the Hollow—people whose houses have disappeared, are falling down, or inhabited by strangers. When I knew them, Annie and Wade lived an hour away by car, but the psychic and cultural distances were great. They experienced little sense of family or community in Pulaski. "I live near people," Wade said, "[who are] within throwing distance of my house [and who] have never been in my house, never visit, and I don't visit them." Annie was living in an apartment building where, she said, "you can look across the street, and they are selling drugs." Annie and Wade led satisfying lives away from the Hollow. But it was not the same. Something was missing.

Of course, they were nostalgic. Of course, they idealized their home and Clover Hollow. And perhaps their adult lives were not as full as they would have liked. Nevertheless, their memories and their sense of place had solid bottom. Something was missing from their present lives. Wade and Annie both felt placeless—displaced in some degree—a feeling I certainly understand. This Clover Hollow home place—this primal place—had shaped the people they had become and given them a sense of meaning and security. It centered their lives. They were repeatedly drawn back. The land was still there for them to walk on. The decaying house brought images to mind of the way they once lived. It also reminded them of change and loss. That original place, nevertheless, lived in their memories and imaginations.

7.

The 1892 Givens Home Place

Soon after I moved to Clover Hollow, I met Caroline Givens and her husband, Bill Vincel. They were warm and generous from the start, and over time, we became good friends and I learned a great deal from them. Caroline was the granddaughter of Joseph Cale Givens, the second of James Stafford Givens's sons to move to Clover Hollow.

In 1878, Cale and his family moved into a small log house at the base of Johns Creek Mountain, across a wide field and slightly above the site where he would build the Givens home place in 1892. His son, Bittle, Caroline's father, took over the log house when he married and expanded it from a two-thirds house into a full three-bay I-plan house. All of his children, except Caroline, were born there. She was born in the 1892 house after the family moved there in 1917. It is this house that tells the story of four generations of Givens and a further story about Clover Hollow.

Caroline lived her entire life in that 1892 house. She attended school in Newport. She helped with farm and housework. She stayed home to care for her parents while her brothers and sister left. She continued to live on the farm after she married. She raised her children there. In Newport, she worked first as a postal clerk and then as postmaster. "I've never known anything else" but the farm and the village, she said. The 1892 house and the farm were her primal place—the place that formed the woman she became.

From seventh grade on, when her mother suffered her first stroke, Caroline, the youngest child, assumed most of the household responsibilities. "After high school," she explained, "my life was nothing but work." Her mother's second stroke left Caroline with the full burden of parental and household care. Mary Givens died in 1943 when Caroline was twenty. Her father kept his health for another seven years, but he was alone. "I wasn't going to go off and leave him there by himself. He couldn't even boil water." Bittle fell ill in 1951 and died in 1955—long past the time when Caroline, now married, and Bill could have chosen a different life.

Caroline's brothers and sister, however, left as soon as they could. The boys went to college. Her sister enrolled in business school. Since both parents had graduated from college, education beyond high school was a family expectation. The brothers went on to professional careers. Ed, the oldest, became an executive with Southern States and a prominent alumnus of Virginia Tech. Lawrence worked for the Department of the Interior and at his death was in charge of Game and Wildlife Refuges and Game Management for the southeastern United States. Jim taught vocational agriculture in high school. Frances attended business school for two years, then married, and moved away. None of them returned to Clover Hollow to farm or live, but most stayed closely attached psychologically and returned to visit the home place whenever they could.

Lawrence "was bound and determined," Caroline explained, "that I was going [to college]. And he went to the bank to borrow money. He had everything fixed up." But she couldn't (or wouldn't) go. "I can't go off and leave my father," she told her brother. "So therefore, I never got a college education. But I don't regret it."

When I knew her, Caroline was in her seventies. Her hair had turned completely white, and she suffered the spinal curvature and pain of osteoporosis. Years of working in the sun had permanently damaged the skin on her hands and arms. She could not do nearly as much as she once did at home, in her gardens, or in the community. She was nevertheless gracious and engaging, often witty and ironic. She smiled frequently as she talked about her life, the Hollow, and Newport. But she was realistic about it all. I sensed only a trace of nostalgia when she described the changes she had seen and expressed anxiety about the future of the farm after they die. She laughed a little stiffly and said, "We won't know about it."

Bill and Caroline were married in 1946 after his discharge from the army. They had met before World War II when he went to Newport from Virginia Polytechnic Institute (VPI) to teach during the 1939–40 school year. Caroline graduated from high school in 1940 and worked briefly as a secretary at the school, but care for her family soon became her whole life. After the war, Bill had a job lined up elsewhere, but he could not accept it. "What are we going to do with Dad?" she would ask. They decided to stay on the farm but only after Bittle, Caroline's father, agreed to leave it to them. "All right, you stay here and take care of me"—Caroline was remembering what he said—"and I will let you have this place." When he died, Bill and Caroline did inherit the farm; they also inherited the debt.

The 1892 Givens home place is a classic T-plan, two-story farmhouse constructed with three bays across and three deep with the central bay, or the "passage"—including entrance hall, stairway, and upstairs hall—common to the wide front facing the road and the extension to the rear. (See ground plan 3, page 30.) A two-story front porch with a gabled roof extends forward from the face of the central bay. Both the lower and upper doorways are framed with a transom and sidelights. The windows facing the main road are centered in each outside bay on both floors. Brick chimneys stand on each gabled side of the house. A third chimney, at the back of the house, served the kitchen fireplace. Porches ran along the length of each side of the rear extension. The one on the west had two levels—the only exception to the symmetry and balance of the whole. It is a beautifully symmetrical design.

The foundation was made with limestone gathered on the farm. The beams were hand hewn from timber cut on the property. The house is framed in oak milled on site. "They moved a sawmill in," Bill explained, "and sawed all the timber for it right here on our

Image 6. 1892 Givens Home Place, ca. 2000. Image by author.

place." The siding was cut from poplars growing on the mountain. The chimneys were made from clay dug out of the cistern. "They made them right here along the fence," Bill told me. "You can still find traces [one hundred years later]. They pressed the bricks, then made the kiln, and fired them in that. A lot of them turned soft and had to be discarded, but they got enough to build three chimneys." Two of the originals were still there. The third, on the north side of the house, had been replaced in the late 1950s after the earthquake of 1957 cracked it beyond repair.

The home place was well designed, decorated rather simply, and soundly built to serve a practical purpose—convenient living for a hardworking farm family. Its basic design satisfied expectations of simplicity, adaptability, practicality, and harmony for farmhouses. It was not conceived primarily as an aesthetic object. Nevertheless, one of the manuals of the day, *A Pocket Manual of Rural Architecture*, advised, "Convenience and comfort are the first requirements of a farmhouse; but there is no reason here, more than in any other sort of residence, why regard should not be had to the beauty of the external features."

Joseph Cale's carpenters made a serviceable and an aesthetically pleasing home. The symmetrical central hall design not only is simple and elegant, but it also made heating and cooling relatively easy. There were fireplaces in every room except the hallways, the dining room, and the bedroom above it. They heated the spaces fairly well in winter. The family did, however, have to pass through a cold hallway and the cold dining room or walk along the side porches to get from the front of the house to the kitchen. "A lot of times the snow was so deep on the porch you couldn't come through it, so you had to come through the hall."

Comfort and convenience are relative terms, and compared to the log dwellings people erected earlier in the century—like Cale's first house and the Lafon houses—the 1892 home place was more practical, comfortable, and attractive. But it was not nearly as comfortable as we now like homes to be. It had no electricity, no central heat, no insulation, and no indoor plumbing. "All I can tell was it was cold," Caroline remembered—especially early in the morning before the kitchen fire began to warm the room.

During warm weather, the house was well ventilated. Air flowed from the front door through the central hall, the dining room, into the kitchen, and out, or from the kitchen to the front, and through front windows and doors. Breezes blew across the rear section of the house, cooling the kitchen and carrying away the heat of the day from the bedrooms above.

I gaze at a photograph of Joseph Cale Givens and his family seated and standing in front of the home place. And I try to imagine the family at home. The photo dates from around 1910. Cale and his wife, Nanny, are sitting out in the yard to the right of the center porches—he in a straight-backed chair, she in a high-backed wicker chair. Mattie, Leslie, Bittle, Emmett, and Vernie are standing on the lower porch. Frank (who became a physician) sits to the

Image 7. The Givens family, ca. 1910. Courtesy of the Givens-Vincel family.

left on the front steps. Berta stands at the right corner post of the porch, and John is sitting in the grass between his father and their dog. It is a formal family photograph. Three of the sons are wearing dress shirts and ties. Two wear coats and hats—Frank's bowler is pulled down over his eyes; Emmett's hat is pushed back showing his forehead and dark hair. Joseph Cale is coatless but is wearing a tie and braces. Nanny wears a long dark dress. Her hair is parted in the middle and pulled back. The three sisters are dressed in white blouses, with puffed sleeves and high necks, and long white skirts. Their hair is pulled up in top knots.

This is the home place as it was constructed, and this is the family that first lived there. The porches show modest but distinct and handsome scrollwork. The corner braces between each post and the portico are cut in a Console pattern (a scroll-shaped bracket). The porch railings are cut in simple Gothic shapes. The roof is covered with wood shingles. The chimneys are nicely grouted and finished. A five-pointed star—like the one on the Lafon house—has been painted in the center of the portico. The four front windows are raised, showing white lace curtains within.

Even though they are all posing, everyone looks comfortable and at home. It is a sunny, late spring day. The trees and shrubs are fully leafed. The dandelions in the grass have gone to seed. The Givenses could have just returned from church on the other side of Big Ridge. The portrait shows a home—a family *place*. It is their dwelling in the world—a relatively convenient, comfortable, and beautiful home place. And it is still a family home place. Caroline and Bill are deceased, but their daughter, Jean, now lives there—the fourth generation to dwell on the farm.

I imagine the home as it was first built and the family's life in it from the photograph. I might be sentimentally dreaming a life for them. Still, this picture and Bill and Caroline's stories have given me some insight into the family's past. From my own experience, I remember the home place well. I spent many satisfying hours there talking with them. There were cold winter days when we sat in the living room warmed by the fire in the restored fireplace. There were mild summer days when we sat on the north side porch or walked around the house and the outbuildings. They talked; I mostly asked questions and listened. I also remember the day Bill and I drove to the Homestead resort in my pickup truck to collect fireworks for the Newport Agricultural Fair. Bill was an enthusiastic amateur historian, so every time we were together he told stories about the Givens family, the valley, and village—about the stores that once served Newport, the old iron furnace along Sinking Creek, the family cemeteries in Clover Hollow, and the covered bridges of Giles County.

During Caroline's youth, the family farmed much as Joseph Cale had when he first moved to the Hollow. She assisted with household chores, as a kind of preparation for her teenage and adult years. She also helped with milking (often outdoors) and making hay. The cows "would just stand anywhere you told them," she explained. You'd "take your bucket and stool and sit down and start milking. When the weather "got so you couldn't stand it, you would drive them into a shed or something." Feeding the animals "was a lot harder [in those days]. The hay was usually stacked in the field. You'd have to go out there in every kind of weather—rain, snow, wind, whatever, and toss the hay off. Then drag it across the field and spread it out for the cows." In summer, when they put up the hay, Caroline's "job was to ride the horse and haul the hay shocks."

"Many times," Ed Givens recalled, "we would take grain—and I would throw a bag on the back of a saddle horse—and take it down to Zell's Mill and have the grain ground for feed or flour. We had no prepared food, and a lot of the soap we used, lye soap, we made it ourselves. We made apple butter, sorghum molasses—get up before daylight and grind all day on the cane. People were very self-supportive."

Bill and Caroline farmed in much the same way during the early years of their marriage—the farm providing most of their food, milk, eggs, vegetables, preservatives, and pork. There was also a modest cash crop economy in the Hollow, and people marketed a wider variety of products than they do now. They would kill turkeys and ship them north in time for Thanksgiving. "We'd run the point of a knife through their mouth into their brain," Bill explained, "twist it and cut the arteries so they'd bleed, take the feathers off, let 'em cool overnight, don't eviscerate them, leave the feet on, and pack 'em in a barrel. You hoped and prayed for cold weather. That was the only refrigeration." Inspectors today "would die in their tracks." The family slaughtered and shipped rabbits in the same way. They'd send crates of chickens as well. "Put 'em in a crate. Put a tag on 'em and take them to the railroad," Bill said.

"About ten days, here comes your check and the crate back." They also marketed extra produce and beef hides. "When you killed a beef, wrap the hide up in a sack, put a tag on it, and send it to Baltimore. They'd buy it."

The Vincels sold milk to Carnation until 1966 when changes in state regulations made the economics of small-scale dairy farming unprofitable. "It wasn't worth while for us little producers to upgrade, so we just quit," Bill said. He also ran an artificial breeding business until all the dairy farms "dried up." Then he began teaching at Giles High School. By this time, they and others were dealing almost exclusively with beef cattle.

These were dramatic changes in farming. The transformation was driven by the modernization of equipment, the economics of agriculture, improved roads and transportation, and new off-the-farm opportunities. "Clover Hollow has gone from diversified [subsistence] farming to practically all beef cattle," Bill told me. Off-the-farm opportunities "started back in the thirties when Mountain Lake [a nearby resort] was being built, and they hired some laborers. It was thirty-nine when Celanese and the powder plant [the Radford Arsenal] opened. That's when it started. Then we got better roads and [Virginia] Tech started expanding." Work off the farm and a less diverse agriculture were the two biggest changes in Clover Hollow, as Bill saw it.

Caroline started working at the post office a few years before her father died and continued, first as a clerk and then as postmaster, for twenty-eight years. Her job provided badly needed cash. It "got our heads above," Caroline said. The post office also gave her a sense of achievement as a capable woman and important figure in the community. She served people with the same generosity and concern she had devoted to caring for her mother and father. She was at the center of village and Hollow life, so she knew everyone and everything. She also "did a little bit of everything"—including weighing babies and babysitting. "Mothers would bring in the babies periodically to check their weight." Some even came by and asked her to "keep" a baby for a while.

Located in a small side room in the old bank building at the center of the village, the post office was not only the source of mail; it was also an information and "service" crossroads. Since there was no home delivery of Sunday newspapers, for example, the papers were dropped off at the post office. "I would get up every Sunday morning," Caroline recounts, "go down there, and put the papers out, so the people could get them. Finally, I had to cut it. It got to be too much, but I did it for a number of years. When you're a postmaster, you really tried to give service."

Caroline left the post office in 1980. She had earned enough credit to retire, and she was weary of the work. Nevertheless, she felt a sense of loss. "I miss the contact with the people. Well, I knew everybody. I was the first person, really, to meet the people who were new to the community because they'd come in there about mail. And I really miss that."

Over time, Caroline and Bill modernized the interior of the 1892 house. Bittle had connected it to Appalachian Power, which came to the Hollow in 1936. He also replaced the wooden shingle roof with tin after the barn burned. But beyond that, he couldn't afford other improvements—nor could Caroline and Bill for a time. "Our pickin's was a little thin for a few years," Caroline said. Starting in the 1950s, however, they began improving the house. They added indoor plumbing, converted the pantry into a bathroom, remodeled the kitchen, installed a furnace, enclosed the porches, blew in insulation, and replaced the windows with insulated ones. Caroline's post office work and later Bill's teaching provided the necessary cash.

They made the house much more comfortable and convenient, but they couldn't afford to restore the historical features of the interior. When the old horsehair plaster deteriorated, for example, they covered the walls with wood grain paneling and the ceilings with tile, rather than replaster and hang new wallpaper. "We had to do it the cheapest way we could," Caroline said.

They did maintain the architectural character of the elegant, symmetrical exterior. When I walked around and photographed the house, it looked like the house Cale built, except for a few details. The star was gone from the front portico. They also preserved the residential and family integrity of the 1892 home place. It continued to be a practical and pleasing farm home. By living there, Caroline and Bill "kept the family together. They all love to come back," Caroline explained.

Their children, Jean and Norman, echoed that. Norm expressed deep emotional attachments to the home place—"closeness" and "ties" were his words. He talked about the beauty of the Hollow and its peacefulness and tranquility. "I miss that," he said. At the time we talked, he was director of marketing for Select Sires, where he was involved in agriculture but far from farming and the home place. Jean, on the other hand, now lives on the farm. After a career in Northern Virginia with GTE, she moved back "home," as she always knew she would. She also returned to try to preserve the farm somehow. "Everybody comes back to visit. It's always been home for everybody," she said. "We are still a close family."

Despite the changes inside the house and on the farm, the Givens home place was the *same place* it had always been—Caroline's own corner of the world, her center and her security. Unlike so many who moved on in the American way—her brother Ed being one—Caroline stayed. She and the place were inseparably intertwined her entire life: "I've never known anything else." The 1892 Givens home place was her destiny, and it shaped in large measure the woman she became.

Besides expressing and enabling a form of rural life, the Givens home speaks also about the past and about continuity and change—about sustaining life. In May 1998, Bill and Caroline sold virtually all of their farm equipment at auction—tractors, balers, hay rakes, a plow, disc, manure spreader, wagons, and many small tools. I watched through the morning as friends, neighbors, and strangers bid, bought, and then left with their purchases. Bill moved about talking with people and checking prices. Caroline stayed inside. She could not bear to watch. The sale ended more than a hundred years of the Joseph Cale Givens family farming in Clover Hollow. However, Jean's return to the farm—her homecoming after years away—sustains the family's life there and maintains continuity in spite of profound change. Four generations of Givens/Vincels have lived there.

———

The Givens home place also speaks about the way we dwell in the world, if I may be philosophic for just a moment. In Norberg-Schulz's terms, the house and the farm manifested in concrete terms the way "man" (his term) dwells in the world. We are grounded, or embedded, in particular places. Our identities, he proposes, depend on belonging to a place. Our identities, in my terms, are formed substantially by that place. And without such a place, we are homeless. As Edward S. Casey proposes, "To lack a primal place is to be 'homeless' indeed, not only in the literal sense of having no permanently sheltering structure but also as being without any effective means of orientation in a complex and confusing world. By late modern times, this world had become increasingly placeless, a matter of mere sites instead of lived places, of sudden displacements rather than of perduring implacements."

When we truly dwell in a place—rather than simply reside somewhere—we identify with it, and we orient ourselves within it. That is, we identify by internalizing its meanings and discovering its "genius loci," as generations of Clover Hollow families intuitively did. We orient ourselves in a place by creating domains with paths, roads, fields, fences, buildings, and communities. Or if we come to an established place, we orient ourselves in that existing domain and identify with it as best we can. Without identification and orientation—without truly dwelling—we are homeless indeed.

The 1892 home place was the primal or primary place for each Givens raised there. They were inextricably intertwined with the house, the farm, the Hollow, the mountains, and Newport. I know that expressly from the example of Caroline, the woman who stayed, the woman who never knew anything else, and from long conversations with her brother, Ed, and with Norman and Jean.

The 1875 Lafon home place speaks about the past and change as well, but it also tells a story of decline and abandonment—perhaps about the death of a history and a place. It is falling apart. The land is no longer cultivated or used. It has gone wild. Nevertheless, the Lafon family—when I knew them—kept alive in their memories and imaginations the experience and meanings of that abandoned house. Their sense of self—their very identities—seemed to depend on the place. It had shaped them deeply. It continued to center their emotional and mental lives—if not their daily lives. The 1875 home place grounded them in spite of the loss of the dwelling and the land as a lived place. That in itself suggests something about the power of place.

8.

_____

Outsiders Fitting In

Three Meadow Mountain told a third story—one about the past, to be sure, but also a story about innovation and the future. The house did not fit into the landscape in ways the old farmhouses did. But at its best, it honored the past and at the same time introduced something new—an architecturally inventive house placed where there had been no dwelling or building before and a way of life that brought something new to the Hollow. At our best, even though we were not part of the valley's history and ancestry, we supplemented the forms of respect and service that had been part of life there for two hundred years. We lived with the land largely as it was. We altered the meadow as little as possible and left the woods intact. We sustained and improved the land as best we could. We extended, I would like to think, but did not displace, the forms of life in the valley.

Using a variety of local plants, Barbara established and cultivated gardens, restoring life and variety to an unused and overgrown meadow. Over time, she created twenty-three beds of flowers and plants, including alyssum, asters, begonias, carnations, chrysanthemums, cornflowers, cosmos, daffodils, dahlias, four-o'clocks, foxglove, impatiens, marigolds, nasturtiums, poppies, roses, rudbeckia, Sweet William, tulips, and much more.

I maintained and worked slowly to restore the pastures as open space and hay fields. I replaced the old yellow John Deere and green 740 with a front loader, blade, and bush hog. With it, I mowed my fields, cleared snow from my road, redressed it in spring, and throughout the growing season, I fought "the stickweed wars," so that within a few years, my neighbor, Graydon Shrader, could cut hay in both the upper and lower meadows. We cleared the wooded areas near the house of brush and noxious weeds to permit natural grasses to grow and wildflowers to flourish. We tried to protect the small wetlands in the pond meadow.

Compared to Doak Lucas's fifty years of farming his land, we very modestly, perhaps insignificantly, maintained, protected, and improved. Doak, after all, sustained the way of life brought here by his ancestors. He was seventh generation in Newport and Clover Hollow. He farmed more than three hundred acres. He kept, at various times, cattle, sheep, dairy cows, and chickens. He grew corn and other grains. He built a life and sustained a family. He possessed, served, and improved the land.

Even so, we were trying to find our way and make a place for ourselves, and I keep coming back to the house as the source. I would begin arriving at Three Meadow Mountain from the moment I turned onto Clover Hollow Road at Sinking Creek. The approaches along the roads, the first sightings of the house, the pause at the foot of the drive, and the ascent to the middle meadow would bring me closer and closer to the entrance—where family, friends, and neighbors typically entered. The long walkway from the edge of the road, the slightly inclined pavement leading to the entry, and the clear glass of the door and side panels drew people across the threshold and into the house. They did not come to steps, a solid door, a blank face. The house welcomed them almost as if there were no barrier. The outside virtually flowed inside. Three Meadow Mountain was not closed in the way the traditional houses were. The difference between outside and inside, however, was not altogether erased. The doorway was framed and set back from the plane of the house. The tile changed from Florentine outside to Mexican inside. There was a threshold. From the edge of the walkway, the solid red wall of the guest room partially obscured the entry, but as people walked forward the angled wall began to embrace them and move them forward. Demarcation, threshold, invitation—it was unmistakable.

I actually enjoyed many different approaches to Three Meadow Mountain. The roads did not change, nor did the location of the house. But I approached in daylight and darkness, sunlight and moonlight, through rain, snow, and fog, under clear skies and cloudy, during spring, summer, fall, and winter, and all the modulations within each season, in the morning and late afternoon. Every day, every moment provided a new approach.

And then, I would arrive at this special place. The character of a place depends on how things are made (their spatial configurations, volumes, and decoration), how people dwell within, and how they fit into their domain. Three Meadow Mountain became our corner of the world—our existential foothold for the time we lived there. It enabled us to enter other domains and return. It served our needs for both society and solitude. It shared, in that sense, a common character with the Givens home place and with what the Lafon home place once was.

The house spoke for us. It extended us into the community. It was symbolically an extension of our hands—an extension, I would also say, of our social bodies and selves toward the place and the people. This notion of the house as a hand—as my hand reaching out to grasp the hands of others—helped me understand and explain my experience. Mark Givens and his words helped me maintain my perspective. Mine are, after all, the hands and the gestures of an outsider.

I had met and talked with Mark early on, but I learned the most from him the day I spent in his farm truck while he moved from farm to farm loading, hauling, stacking, and storing large, round bales of hay. The truck was a large flatbed International diesel—mostly anyway. Mark had cobbled it together from several others.

We were heading to Sinking Creek Valley to pick up the last eleven bales from an eighteen-acre bottomland field. Mark turned off Route 42 onto a gravel road paralleling the creek, drove past three old farmhouses, twisted around a barn, and then drove

into the large open field. He climbed onto his green John Deere
tractor, speared each bale with the pointed front-end attachment,
and loaded them one by one, shaking the truck with each bale
dropped onto the flatbed. He drove back to Clover Hollow where
he mounted another tractor, speared each bale again, and stacked
them one by one in a barn he leases.

Along the way, he talked about the farm and his unexpected
fate. When his father died at age sixty, Mark gave up a sales career
and returned to the farm. "I had it made until he died. I haven't
had it made since. I was saving money. I enjoyed the job. I loved
dealing with people," Mark said. He was living on the farm—
enjoying its beauty and valuing the sense of place and home it
provided—"I think it's the greatest place I've ever been in my
life"—while he was employed elsewhere in a job he loved. He still
did not fully understand why he went back. "It doesn't really sort
out. It was a bad time, a real bad time. Everything was in such a
blur. You look back on it as kind of wandering around in the dark,"
he said. But as he talked, it became clearer.

He loved and admired his father, Daniel, and as a boy and young
man he couldn't get enough of being with him and "learning stuff.
If he had been hauling hay this morning, you'd have to beat me
out of the truck to keep me from bein' here," Mark said. He liked
farming—at least on its good days. Moreover, he didn't want to see
the work there come to a stop. He is fourth generation on the farm.
Mark and his family live in the house his great-grand father, Floyd
Harvey Givens, built in 1916. "When you're where your great-
grand parents were, you feel some responsibility to keep it looking
half decent," he explained.

It was not easy, however. Mark owns about six hundred acres
of one of the "hardest" farms in Giles County, his cousin once
observed. It is rocky and mountainous with limited bottomland.
Mark leases land elsewhere in Giles and two other counties. He
raises cattle and sheep and grows and sells hay. He works twelve-
to fourteen-hour days just to keep up. "This kind of place eats

people," he said. "There's a lot of strain to it if you take it serious."
The work is stressful, tiring, relentless, and marginal financially.

After Mark stacked and stored the bales, we walked up to his house and sat on the porch—now talking about property rights. Farming is so demanding that Mark fights every perceived or imagined threat from outsiders. The U.S. Forest Service, national parks, and the Appalachian Trail Conservancy "just scare the hell out of me," he said. One of them or some other entity might seize part of his land for trails, a parking lot, or some other public use. "You need to be able to do with your land what you want to up to a point," Mark said. "I'm not opposed to zoning. We need some protection, but a bunch of do-gooders don't need to come in and tell you what you're going to do after that."

And that is when he turned to me and talked about my house in Clover Hollow. "I don't see you all as invaders. You all haven't tried to change anything or bothered anyone." I might have been acceptable—even liked, but in reality: "You and me don't make any difference to one another. It's not like that with people around here I've known for a long time." Mark and I did not make the kind of difference to one another that kinship and old friendships make.

The day ended. I headed for my car and drove back to Three Meadow Mountain with a clear sense that I was *in* this place, and accepted, but surely not *of* it. I also left with his last words about farming: "In a way, I don't see why anybody wants to do this to themselves: work all these hours and never be able to plan to be done, to fight the weather. It's a dumb existence."

I imagine an improbable conversation between Jim Cox and Mark Givens. They would talk about coming back—at different times of their lives, to be sure. They would talk about how hard farm work is and their struggles to sustain their home places. They would explore how each of them thinks about their land, their ancestors, the past, and, I would like to think, about the power of place. Maybe. It would be an interesting dialogue.

Mark had put things in perspective. At about the same time, a friend gave a political meaning to our move to the valley. We had made, she said, "the quintessential bourgeois move." We built an expensive house in the country. We surrounded ourselves with eighty acres. We withdrew, to some extent, from others and presumably disengaged from serious social problems. We moved to a valley and community that was virtually all white and disconnected from larger social and economic issues. We lived in the big house, high on the mountainside, and looked down on other people and their land and homes. We looked down on the class struggle, as one academic wag put it. Well, perhaps, but at best that was just partially true.

I could not deny Mark's sense of my difference—I was an outsider. Nor could I reject out of hand the class issue my friend raised. I would be foolish not to acknowledge the inescapable symbolic meanings of this large, professionally designed home. Nevertheless, just as initial acceptance was not the whole story, difference and detachment were not either. We had opened our house to our Clover Hollow neighbors. We participated in major community projects. We were writing and making the people and the Hollow known to others. We were, in fact, trying to make a place for ourselves and connect with others.

Doris Link described the gap between been-heres and come-heres even more pointedly than Mark had. "The people that have always lived here have the kind of connectedness that nobody else can have. They've always known the same people and belonged to the same churches. There is a difference if you've been raised in a place forever. There's just a difference in how you feel about the people you've always known and the new people." She is a seventh-generation resident and lives near Clover Hollow in Plow Screw on land that had been in her family since the late eighteenth century. The people of Plow Screw took pride in their family histories and share a special sense of identity.

As strongly as Doris felt about her ancestors and her land,

she also understood the value of collaborating with newcomers to sustain tradition, confront threats, and celebrate the past and present of the Newport area. Outsiders brought talents and knowledge that could serve insider interests as well as their own. Without the leadership, for example, of certain newcomers, the community would probably not have come together to fight construction of a 765 kV power line through the area or write a successful proposal for the Greater Newport Rural Historic District.

There was a community of interest among newcomers and natives that focused on issues and specific projects to preserve and improve the place we had all—at one time or another—been given to care for. Insiders and outsiders achieved real—if momentary and selective—bonding in these situations. "The underlying thing is a focus or cause or whatever you want to call it," Doris explained, "a thing where we all think and feel about the same, even though some's been here a long time and some's been here a short while. You put aside differences or personal preferences to concentrate on getting the job done."

In 1991, American Electric Power (AEP) filed an application to construct a 115-mile 765 kV power line from Wyoming County, West Virginia, across the Alleghenies to Cloverdale, Virginia, near Roanoke. Several of the proposed corridors ran near or through Newport and Clover Hollow. The power line threatened the environment, communities, and culture of the region. Shortly after the AEP filing, opposition groups began forming in both West Virginia and Virginia. Our local group, Citizens Organized to Protect the Environment (COPE), came together in early 1992 as a coalition of natives and newcomers. It fought the line for almost a decade, and in the end succeeded. I served on the Steering Committee for a short time.

COPE fought the line through direct political and legal action, and that required knowledgeable activists. This is precisely the kind of experience and knowledge that newcomers could provide and that insiders were less likely to have. As president, Doris Link quite happily joined with newcomers to preserve the land and way of life she so dearly loved, but that did not change her sense of difference. The power line was simply one of those important things "where we all think and feel about the same."

In 2000, the Virginia Department of Historic Resources designated the rural areas around Newport as the Greater Newport Rural Historic District. The district is also listed on the National Historic Register. Dorothy Domermuth, a newcomer to Plow Screw and Clover Hollow, and Nancy Kate Givens, a native of the Hollow, led a working group of roughly twenty people evenly divided between natives and newcomers. The whole process took six years and the collaboration, once again, of insiders and outsiders. "There was no conflict," Dorothy said. "I've never seen a group that worked so well together. And it was old and new. It gave us a sense of belongingness."

The power line threat started it all. "My original interest," Dorothy explained, "was not for the historic part so much as just fighting the power line." Even though the project made Doris Link "aware of how deeply I felt about this place," she agreed with the group's primary purpose. "We were advised that we would be a stumbling block in the way of the power company." Later on, the group recognized the intrinsic historic value of the project. "We are a historic district. We have a historic way of life. The activities that go on here could have gone on a hundred years ago in somewhat the same manner." Besides being a "thing where we all think and feel about the same," the historic district project is also an example of the way outsiders can help insiders discover what they have.

Jimmy Lucas, Doak's son and then a resident of the Hollow, said that newcomers had actually made the community closer. "It's quite possibly closer because of people like yourself, people who have been involved. You are more interested in our heritage than we have been because we take it for granted," he said. If Newport and Clover Hollow value history and tradition in the future, Jimmy continued, "it will be primarily from people like yourself instead of people who grew up here."

The Newport Agricultural Fair, more than any other local event or celebration, symbolizes the continuity of the past with the present and the power of place for local people. Established in 1936, it enacts what the historic district recognizes and honors. It is a small country fair organized and run by local people. It runs for two days every August, opening Friday at noon and closing late Saturday night with music, dancing, and fireworks. The fair features a wide range of exhibits, food tents, games for children, competitions for adults, and entertainment. It all takes place at the old school, now a community center, in the adjacent rescue squad building and on a few acres next to the school.

The fair has become a community homecoming attended mainly by people with some present or past connection to Newport. It celebrates both the present, by representing what people still do and care about, and the past, by speaking for the community's agricultural heritage and its sense of continuity and identity. It preserves a heritage and helps sustain a way of life.

For most of its history, the fair was organized and run by insiders. Over the years, however, the board of directors did include a few newcomers with strong commitments to sustaining and improving the community. I was one for a few years, and I understood my place. I was assisting, not leading. I was helpful but not necessary. I participated on behalf of the natives and their vision of things and to preserve the place I had come to live, as other newcomers had with COPE and the historic district group. Even with this most traditional community event, insiders and outsiders collaborated, but the natives clearly led and dominated.

"He's putting down roots," an amused Norm Baker said years ago. We were at Alexander's restaurant in Roanoke. I'd just told him that Three Meadow Mountain was under construction. We had been colleagues and friends in Ohio. Norm knew about my peripatetic past—Delaware, Columbus, the Air Force, Bloomington, Athens, Greencastle, East Lansing, Oxford, and finally Blacksburg. He spoke casually and by "roots," he simply meant staying in a place and not moving on—right away, at least. But in a sense, he said more than he meant. Or, perhaps better, his words took on a deeper meaning over time as I grasped more fully what had happened and began to understand my own past and the power of place in my life.

It took time for me to understand that the "roots" I was "putting down" were enabling me to make a place and join a community like none I had known since my hometown. I was still a come-here, but people accepted and valued me—not only because Barbara and I hosted friends and neighbors and joined in community action, but also because I was writing about families that had been there for generations and perhaps enabling them to value themselves in a new way. For me, writing seemed like one more way of protecting and articulating the place I'd been given.

It took even more time for me to realize that what (I thought) was missing in my life had always been there—inside me, as well as back there in time and place. Clover Hollow transformed my sense of my past, my hometown, and myself. I had met people in the Hollow like none I had ever known—people with a rich sense of place and past and with a firm sense of community and shared life. I had been trying for years to make my life intelligible, but I still thought of myself as a man always leaving his pasts. I had been living essentially a placeless life—although with a certain professional and personal continuity. Clover Hollow and Newport, however, interrupted the neat, progressive, academic life I had led and forced me to see myself in a different way.

At that point, I turned all I knew about the valley and village on myself in order to recapture my own heritage, so richly embedded in places. I experienced no "aha!" moment—no epiphany. It just happened slowly over time and rather late in life. At some point early in that process, Jim Cox had walked into my office.

I left Three Meadow Mountain and Virginia because I had other lives to lead and other places to go—none, however, quite like my life in Clover Hollow and none so profound in its influence. Barbara and I had transformed open space into a place for ourselves. We had built a house that paid homage to the past architecturally. We were also telling stories about the people and the places of Clover Hollow and Newport. I was publishing essays. Barbara wrote and produced "The Newport Play" for the annual Newport Harvest Festival. We were reaching into the past, writing about the present, and participating in community action. We were also reaching beyond the village and Hollow into the future.

We had tried to make Three Meadow Mountain our place in a valley and among families with one and two hundred–year histories. That place—Clover Hollow in all its complexity—became a site of discovery for me, but in the end, I realized that my deepest roots could not in fact be there. I remained a come-here, and I knew I would always be.

So once again, I moved on, drawn by those other lives—but now with an emerging understanding of the way my hometown had made me the person I had become. My life in Clover Hollow led me back in time and memory to that hometown and to my father who had died at age forty-six.

9.

_____

Interlude

Three Meadow Mountain lives in my memory as a place of delight and discovery. I used to stand on the upper porch in sunshine, rain, and snow, on mild as well as cold days, gazing toward the Farriers' red barn in the bottom of the Hollow, scanning the scene from Kelly Knob on my left to Clover Hollow Mountain before me, and then down the valley toward Marty's at the foot of the Hollow; hearing the bellowing of cattle, the steady beat of tractor engines in the distance, the hum of automobile tires on the paved road, the wind in the trees, rain on the porch roof—all the music of this valley. I would listen and watch as if it were all mine, but of course it was not.

Occasionally on a summer evening, after a rain shower had passed and then hovered over Kelly Knob, the descending sun created a magnificent rainbow that arched over the Knob.

On cold fall mornings, a dense cloud sometimes dropped down into the valley covering everything below Three Meadow Mountain. The sun would rise over Kelly Knob in a cloudless sky. Standing in that early morning light, looking up at the blue sky aglow with sunlight, I would feel above the world. When I looked straight out over the valley, I could see the ridge of Clover Hollow Mountain above the fog, a cluster of treetops on Big Ridge, and the tip of Spruce Run Mountain in the far distance. The fog covered everything below me.

I also recall my dear friends and neighbors—Bill and Caroline, now dead—others I've lost touch with in my new life. Unknown to them, they made such a difference in the way I think and live.

Clover Hollow meant so much to Barbara that she asked that I bury her remains in the Clover Hollow Cemetery. She died six years after we left.

But enough! I'm on the verge of a usual mistake—an idealized Hollow, pulling me back into nostalgic recollection or pointless mourning. It was difficult to leave emotionally. And in memory, I have not left altogether. But leave we did.

It is time to move on, but let me pause for a moment and review my underlying assumptions. I am writing about place—a complex concept and condition. Place refers not only to geographic and constructed spaces—location, topography, landscape, and buildings. The term refers as well to psychological, social, cultural, and remembered forces active in every place. Places are at once fixed and also dynamic—as Edward S. Casey has so clearly established. That is, places are events. Places happen. When Jim Cox returned to Brookside Farm, he returned to his primal place, which in certain ways had not changed. He also returned to a place that had indeed changed over time and continued to do so every day he lived there.

I am also writing about people *in* places—and especially about individuals in place who were critical to my own discoveries. Self and place, people and places, are thoroughly intertwined. People dwell in places; places dwell in people.

The primary places where we live and have lived—home and town—and influential places, like schools, condition our acts, thoughts, and way of life. Whatever surrounds us modifies our behavior and in our formative years shapes our identity. When I write about my hometown—my primal home, the schools, and the downtown—I am writing about places that conditioned the way I live and shaped who I am—just as Brookside Farm influenced Jim Cox, the 1892 Givens home and Clover Hollow did for Caroline Givens, and the Lafon home place did for Wade and Annie.

Home

10.

———

A Boy from Columbus.
A Man *of* Delaware, Ohio

Not until I understood Jim Cox's rich heritage and his sense of place did I begin thinking about my father's life and heritage and not simply his death and my loss. So after I left Clover Hollow, I set out searching for Ervin—the man before and beyond father. I sought the living Ervin—not the emaciated body of his last days consumed by cancer but the man who lived and loved, a husband, lover, friend to many, an educator, and community leader—a man of Delaware, Ohio, the city he loved and served

This man, I would discover, was shaped deeply by the South Side of Columbus, Ohio, where he lived from infancy until he was a teenager with his German grandparents in a tough, working class part of the city. His twenty-four-year-old mother died before his first birthday. His father soon left, effectively abandoning him. In his first home, his primal place, Ervin learned love from Mama and Papa Döerflein and deep loss from the absence of both parents. In that place and in the schools he attended, he learned to survive and then succeed. Later as a man, he made a place for himself and his family in Delaware, Ohio. My search led me back not only to him but to my hometown. Searching for Ervin was my first turn toward the place and heritage I thought I did not have.

My father's death at age forty-six possessed me for years. When I wasn't thinking expressly about it, his death was there, nevertheless. I had wondered if I would live past his age; I felt relieved when I did; but even then, I continued to dwell on him and my loss. I couldn't escape. It was the way he died—his suffering, my mother's distress

Image 8. Ervin Carlisle. Author's personal archive.

and despair, my shock and confusion—that so possessed me. Cancer killed him. It wracked and wasted, and then it killed him.

Near the end, when he had weakened so, I carried him to the bathroom and helped him sit on the toilet and then shower. He was so light, so emaciated. But soon even that became too much, so my mother cleaned him as he lay in bed.

Near the end, he could not pass the night without a shot. I slept across the upstairs hallway with a long string attached to my wrist. He held the other end. When his pain became acute, he pulled the string. I had to drive a thick, two-inch long needle deep into what little muscle remained to relieve him of his pain. His hips, buttocks, and thighs were badly bruised and discolored from the frequent injections. We hated inflicting such pain and injury. Without morphine, however, his suffering would have been unbearable.

Near the end, when he swallowed water, it sounded hollow and metallic, like water draining through metal pipes. A few days before he died, he whispered, "I am so tired of all this pain."

At the end, he did not go gently. He died thrashing in his bed, fear in his eyes, fighting, perhaps, to stay with my mother and me for one more minute—or simply struggling against the darkness of death. I held his hand and watched him go. The wild thrashing ended. His life left him with a dying rattle. He was quiet. This is the dying and the death I could not escape until Clover Hollow and Jim Cox showed me a way. I followed many roads on my journey home.

When my mother remarried, she passed on to me a miscellaneous family archive in a large plastic storage box. I moved it from place to place over the decades—rarely, if ever, looking at it but not able to dispose of it. Then in Virginia I did finally open it. The pages of Erv's high school and college scrapbook were dry, brittle, and flaking. Every time I opened it, a few more pieces fell away and scattered on my table or in my lap.

The box also contained photographs, newspaper clippings, a few *South High Optics* (his high school newspaper), birth, marriage, and death certificates, file cards of jokes, a handful of letters from Ervin, hundreds of letters written to my mother upon his death, and a handwritten fragment starting to explain his feelings when he learned he was terminal. "When I was told that I might not live very long, the first thing I did was thank God I'd had the foresight to plan for such a contingency. The next reactions seem to come in thousands. . . . All levels, types, and extremes of emotions." That is all. I found nothing more about his feelings. The fragment then turned from him to his family. He had scratched out a draft of financial instructions for Winnie, his wife and my mother. So typical of his concern for others.

Incomplete and accidental as it is—Ervin was living his life and not recording it—the archive triggered my memory and imagination. It told me things I did not know. It immersed me in his life whenever I opened it.

Ervin sat for hours at the dining room table, cutting and splicing short 8 mm film reels into longer ones. He worked carefully and happily, transforming fragments into somewhat more orderly images of his family. He used an old, rudimentary technology—running film through a manual viewer and then cutting and pasting by hand. Hour after hour. He was creating a visual record of the family he and my mother had made together—the kind of life he had not known ever before.

The films were faded. The colors blurred into one another. Very little seemed in focus. I was watching moving images of my family and friends. Some looked familiar. Others I did not remember at all. I was watching, but I was also there—not just *in* the movies as a boy and youth, but somehow there and looking on. I was standing beside the action—beside the man holding the camera. I looked to my side and tried to see him. I almost could.

The home movies last less than two hours and cover only ten years. They are fragments of a life that add to memory. I watched Christmas at our two homes on Lincoln Street and at my maternal grandparents in Gary, Indiana. Erv's camera pans over lighted Christmas trees, the gifts underneath, along the "Merry Xmas" string of lighted bells and the pine boughs on the mantle, toward lighted fireplaces, and back and forth over the holiday dinner table and my grandparents, aunts, uncles, cousins, and my younger self and my mother and sister. All this was so dear to him.

In another sequence, we're sledding down the long, sloping hill behind our house at 208 West Lincoln—down, then back up, dragging our sleds behind us, down, up, over and over. At one point, I'm sledding toward the camera, my face raised toward it, then underneath, and evidently through the arch formed by my father's legs. There were boyhood friends I'd not seen in over fifty years—David, Paul, Terry, Bernard, Connie, Judy.

I am fascinated by the way film, photographic images, and memory flow together—pictures creating memories, memories explaining pictures, images prompting related but unrecorded recollection. After so much time, it becomes almost impossible to separate history, image, memory. I do know when I'm concentrating on a film or photograph and when I'm remembering something without an image to guide me. Nevertheless, the distinctions are soft and fluid. It is complicated further because of the way memory and invention are so intertwined. It's complicated still further by the way invention sometimes tells the truth and sometimes simply expresses a desire we call memory.

I sat for hours and days in the dimly lighted microfilm reading room at the Ohio Historical Society. Except for the clicking and whirring of the machines as readers advanced films or the high hums of fast forward or rewind, and then the flap, flap as a film released from the wind up reel, the room was mostly quiet. Occasionally, someone spoke—a question about operating a machine or how to find a record, readers talking about what they were finding or could not find, expressing excitement or disappointment. I was reading through twenty-three years of my hometown daily newspaper, the *Delaware Gazette*—week by week, month after month, year by year—from the time Ervin graduated from college until he died. He had been a high school coach, then principal at age twenty-six, superintendent after that, and later civic leader. News about him appeared often.

Those microfilms and the blurred images of home movies somehow put me there in the moment with Ervin, observing as events in his life unfolded.

I spent hours in the archives at Ohio Wesleyan University reading the *Ohio Wesleyan Transcript*—following the "midget" quarterback's college football career, game by game, score by score, injury by injury. I had known about the Michigan game for as long as I could remember. In 1928, Ohio Wesleyan beat Michigan 17-7 in Ann Arbor—but I knew no details, nor did I know anything about the rest of the season when they also beat Syracuse. "Michigan was powerless to stop Ohio Wesleyan's vicious attack and fierce defense," one newspaper reported. Ervie's picture was in the paper the next day, so I knew he'd played and probably played well. I remember one football picture, particularly—Ervie, number three, is driving into the center of the University of Dayton line, the ball in his right hand, his knees bent, running low, cutting right, his left arm raised to stiff-arm a tackler.

I followed the football and basketball seasons when he was coaching at Delaware High School. I'd known nothing about them. Each was an adventure as I read about the games, not knowing beforehand whether Ervie's teams would win or lose. At times I was afraid to read on—had "we" won or lost?

After reading years of the *Gazette* and following Ervin's progress in the schools, I knew it was coming. He would resign from a system he had led for five years as superintendent and served for fourteen altogether. But I had forgotten. I was reading the newspaper microfilm so intently—caught up in the details of his school and community work and the World War II news. "Fifth Army Is Near La Spezia" read the headline next to "Ervin F. Carlisle Will Leave School System." The resignation surprised me. I felt hurt, angry—a catch in my throat, tears in my eyes. Moments before I was thinking how rich and valuable his school and community service had been. He's leaving? How could people let that happen?

In late May 1945, the student body of Willis High School held a special assembly to honor Ervin. Several students wrote and then performed a "History of Ervin Carlisle"—thirty minutes of "affectionate ribbing," as reported by the *Delaware Gazette*, starting with his birth (a fifteen-month-old baby appeared as the infant Ervin) "whose advent set the world agog." At the end of the seven ages of Ervin Carlisle, four ghosts arose from their graves, and in close harmony praised him, "the man who insured them." The students presented him with a one hundred dollar bill, a check from the FFA (Future Farmers of America) for buying a movie projector, and a movie screen from the boy's athletic association. The projector disappeared long ago, but I moved that screen from house to house and state to state for many years after his death.

Erv knew he'd have trouble speaking at the assembly, so he prepared. With tears in his eyes, he held up large signs reading, "Thank You; Good Luck; I assure you I'll be interested in you as long as you live; I'll be in the life insurance business." I was touched, again, as I read about him in the *Gazette*. I had been there, at age ten, and saw him on stage holding up the signs trying to smile and not cry.

My search took me to places in Columbus, Ohio, his hometown. I drove by the Döerflein house on the South Side of Columbus—Ervin's primal place, the home that influenced him deeply. It had been so changed by additions, siding, and a picture window that I could not recognize the house I vaguely remembered. I went to his high school—Columbus South—a new school in his time. I walked toward the still impressive facade of the building, through its main doors, and along the hallways. Ervin must have entered through those same doors, walked the same halls, sat in the class-rooms they led to, played basketball on the floor that shared a stage with the auditorium, and played football on McHaffey Field behind the school.

The auditorium was dark and quiet, but I could see the old basketball floor beyond the stage and the auditorium chairs. I could almost hear the ball bouncing, the players running, shoes squeaking as they cut, spectators cheering. At the school's trophy case, I studied the photograph of Ervie's 1927 central district championship team. Its seventy-five-year-old trophy was still displayed—a large silver basketball (dented on the top) resting on a black wooden base. In the main office, I looked at Ervin's barely readable transcripts. In the school's archives I read the student newspaper, the *South High Optics,* from Ervin's time when he was a sports columnist and editor.

I searched for records at the Columbus Country Club where Ervin caddied and then after high school, served as caddy master, introducing many innovations into the lives of the caddies—a club, a newspaper, a governance system, a lecture series about character in sport. I found a few old *Caddy News* at the club, but all of its records from that time had been destroyed years before in a terrible fire. Even so, I could tell that his intelligence, imagination, and gift for leadership were beginning to focus on young people.

I read years of school board minutes and paged through year-books at the Delaware City Schools administrative offices—housed in what was my old 1869 elementary school. At the Ohio Historical Society, I not only read microfilms but also looked through city directories, birth and death records, and photo files. Places and places, record upon record, and there were even more as I searched and gradually found my way back.

I met with relatives from Columbus—Ervin's cousins, Bob Stith, Paul Stith, and Dick Carlisle, and with his half-sister, my aunt Jane. I talked with Delaware friends, former students and players of Erv's, school board and civic colleagues, local professional people, and classmates and cousins of mine. Even though most of his closest friends and many townspeople of his generation had died by the time I started searching, a clear portrait of a good man was emerging.

Bob Stith told me about their grandparents, Ervin's life as a boy and youth, and the neighborhood on the South Side where they lived. It was largely a German immigrant and working-class community and, indeed, a tough part of town. It was mixed ethnically and racially. Not far away were areas of appalling poverty and dilapidation. (I remember my father's stories about his frequent after-school fights.) Grandpa Döerflein worked at a shoe factory. Ervin took his first job at age nine loading bread trucks at a nearby bakery. The family was barely getting by financially.

Bob regarded his cousin as a kind of older brother. They had lived together for a year or two at the Döerflein's. Ervin took Bob to his first day of school at Southwood Elementary. He took him along on the bus to high school football games. He let Bob ride with him on his bread truck route—the bakery job he had as a teenager. Bob cherished these memories and returned to them over and over, affirming his great affection for his cousin.

After reading the transcripts of our conversations, Bob Stith asked, "What's missing from all of this?" I waited. "His father. He was never there. He was never around until your dad became a very popular guy around the community"—a well-known high school athlete. Even my aunt—a woman devoted to her father—affirmed that Ervin and Herbert were not close. "They were away from one another during his formative years," she told me.

Even with the love of the Döerfleins, the boy must have felt a deep sense of loss and emptiness. I was searching for a man who himself was searching for the life and family he never quite had as a child and youth. We were both searching for his life—he in prospect and I in retrospect.

"Football made all of the difference. That was the thing," Ervin's cousin, Dick Carlisle, told me. It made Ervin's later life possible. It allowed him to be aggressive and rough, even violent, in a mainly constructive and competitive way—a socially acceptable way. He was "the toughest football player I ever knew. . . . He wasn't very big, but he was tough"—Dick was paraphrasing a local sports-caster. Football also made him part of a team that shared a purpose and worked together in a coordinated way to compete and win. Football gave him confidence and provided recognition. Ervin became somebody for himself and for the public. It helped fill the emptiness. The game also helped him develop his sense of self and his purpose in life. He was repeatedly described in *The Ohio Wesleyan Transcript* as one of the greatest field generals, "one of the brainiest quarterbacks," a "masterful" team leader.

Although many of Ervin's closest friends and school and civic colleagues had died when I started my search, I did find people who had known him well. Bob May and Gene Peebles were students when Erv was high school principal. Gene later served on the Board of Education with him. Both tried hard to recall details, but it had been fifty years since Erv's death. They were clear, nevertheless, about the kind of man he was.

He was fair and open, Bob recalled, a good principal who encouraged him to do better. "I've been looking at you," Bob remembered him saying, "and I think you could make something of yourself, if you'd just try." He was also a man who "wouldn't take crap off" the self-styled "tough guys," said Bob. "He was one of those people you had to look up to. He kind of demanded it, and we gave it to him. He looked you right in the eye, and you knew he meant business." Both Gene and Bob talked about his "presence"—Erv's warmth, sincerity, generosity, his firmness and integrity, his way of encouraging people and bringing them together on difficult issues.

"Erv always wanted to give people a break," Gene said. "That's good—a good part of a person. If a person makes a mistake, give them another chance. There's something about the way you treat athletics and treat the people [the players] and the way you treat anybody." Gene went on, "You can't get other people to like you unless you like yourself, and he was a man happy with himself. Confident. The people who came from that kind of background are the ones who excelled." Gene was speaking about Erv as a coach and a leader; he might also have been thinking about Erv's relatively deprived, working-class upbringing in Columbus, Ohio.

Harry Humes and Erv had been close friends, as well as school board and community project colleagues. He was in his mid-nineties when we last talked and has since died. On my first visit, Harry showed me the red vest sweater Erv had given him more than fifty years before. He was proud of the gift and their friendship. He described my father as a remarkable public figure and community leader.

As a school board member, Ervin helped resolve serious race issues before the board regarding new school construction and school boundaries. The North End of Delaware was all white, and the South End mainly black. Both needed new elementary schools. There had been some sentiment to do North School first. In the end, the board decided to start both buildings at approximately the same time. It also decided it would not base school boundaries on race—even though some white parents did not want their children assigned to South School. Ervin's ability, Harry said, to placate people and find the positive and common elements in any controversy helped defuse the tension and resolve the issues. "Look at it this way," he typically would say.

The two men served the city well through the last eight years of Erv's life. Ervin was elected president of the Delaware Development Corporation (DDC), an entity he helped found to encourage industrial and business enterprise in Delaware. Harry was also part of the group. The DDC concentrated on Delaware's inadequate housing for postwar development. It also helped bring new industries to town.

Ervin was also elected chair of the City Charter Commission, a group established to transform Delaware city government. Harry was elected secretary. After months of work on a charter to change the city to a city-manager-council form of government, the commission presented the charter to the mayor, council, and the people of Delaware. It then mounted a vigorous campaign advocating adoption. The charter was subsequently approved.

Both men loved their city. Ervin did not live to see it change— and change dramatically. Harry did live to see it and believed a lot had been lost, especially the friendly, small, and trusting town

Delaware once was. "Those were the days that you knew every-body. You probably knew a lot of people, and your dad knew everyone."

Delaware has more than quadrupled its population since 1950. The once self-sufficient downtown is long gone. The current civic leadership is determined to find a new identity and prosperity for its downtown. The old sense of connection however, has disap-peared, as Delaware grows and becomes more of a bedroom com-munity for Columbus. "I don't know that I know one business owner," Harry once said. And he was not alone. High school class-mates I talked to some years ago not only felt the losses, but they felt estranged—excluded, really—from this new Delaware. It had become an unfamiliar and more expensive town for them to live in.

Harry asked me one day about the differences between old and new Delaware: "What do you think of it?" "Well," I said and paused. I didn't want to be sentimental or nostalgic. "In a sense, Harry, this is what you, Erv, and other civic leaders set in motion. It might not be what you envisioned or even wanted, but you set the stage for major growth and development." "Oh," he said and looked at me as if he hadn't thought about it that way.

My cousins Dave Lucas and Bob Martin knew my father well. Dave loved visiting the Carlisle household. "It was a happy house. We'd just move right in like it was our house. It was so easy. The kids seemed free. There weren't strict rules being shouted at them—don't do this, don't do that. He was one of the most unself-ish, nice[st] guys I ever remember. He took care of you. He was the ultimate host."

Bob Martin liked his "Uncle Erv" enormously. His own father was strict and punitive—a harsh and angry man, it seemed to me. "I remember feeling early on," Bob said, "before I became a teenager, that I envied you because you had a father like Erv, and I didn't." During college, Bob lived at my parents' home for two years. He talked about their generosity and their warm social life. "Your dad was always at his best hosting. He loved it." When I asked Bob, finally, what it was like living with my parents, he said simply, "I was very fond of your mom and dad. It was such a welcome change to be in the Carlisle family. It really became home for me. I've been eternally grateful to [them]."

Fred McKinley, a friend, fraternity brother and OWU student shared a room with Bob for a year. It was one of the most important years in his life when he became "a small part of the Carlisle family." Erv's surgery and terminal prognosis occurred that year. "I found myself living in a household where a very young man was living without any hope of recovery. The situation was handled with absolute dignity, and after a while I became accustomed to the routine. As time went on, I gathered enough courage to enter his room and give an occasional back rub. . . . I would not trade the time spent in this household for any time in my life. I really started to grow up, and I will always be grateful. Erv and Winnie were really good to me." He did not use the words, but I believe Fred experienced warmth and generosity. He observed love and devotion under extreme conditions, and he learned about courage and resolve.

Jack McKinnie, my high school friend and college roommate, wrote about his gratitude to Ervin for helping him attend college. Jack starred in basketball at Willis High School and wanted to attend Ohio Wesleyan and play there. But his family could not afford it. So Ervin appealed to local business and community friends to support four years of tuition for Jack. "I was then and continue to be very grateful for that support." In the 1990s, Jack initiated a Delaware County Scholarship at Ohio Wesleyan. "Each year, one student from the city or county is selected to receive this scholarship. That is my way of continuing to say thank you to the Carlisle family and the others involved." He still talks about this every time I see him.

Of Ervin's closest friends, only Louise Sell was still living and able to talk to me. I missed Paul, her husband, by only a few months. I had known them well as neighbors to my mother, adult friends of mine, and part of the Saturday Night Crowd, a group of ten couples (including my parents) who gathered together socially almost every Saturday for decades.

These men and women had known one another through business or school connections before they began socializing. Over time, the "Crowd" made them into lifelong friends. Along with Hank Thomson, Ervin was one of the oldest. He had coached others in high school football and basketball. He hired another as Willis football coach in 1942. Hank and Erv became close friends when Hank was writing sports stories for the *Delaware Gazette* about Erv's high school teams.

The Thomsons and Carlisles were especially close. I called Hank "Uncle Hank" and his wife, Lillian, "Auntie Toots." The two men were like brothers—a term Hank used in a *Gazette* editorial upon Erv's death: "The management of this newspaper has watched and applauded the career of Erv Carlisle throughout his adult years. He has been a real friend and—to one of us—more like a brother." Less than three months before Erv died, Hank wrote about their friendship—not easy for men who rarely expressed affection for one another but felt it intensely. "I seriously doubt that many people in this country have had the kind of friendship we have enjoyed for over twenty years. It's something for which we can all be proud. . . . One of the greatest honors of my life, Erv, is to have known you as I do. I thank God you selected Delaware as your home."

On Saturday nights, the men and women dressed up—the women in dresses, stockings, heels, and jewelry—the men in jackets and ties. The "girls" usually played poker at the dining room table. Much more composed than the men, they did not yell or throw cards and chips at one another as the men might have. The men sat in the kitchen or a rec room, joking, arguing, teasing, telling the same stories over and over and loving them every time, and at some houses playing Ping Pong. "The men would yell a lot at each other," Louise explained, "about sports, politics, the town—you name it." They often shed their jackets during their loud arguments and ping-pong games. Near the end of the evening, husbands and wives got together for snacks or a late supper.

The men's doubles ping-pong represents perfectly their boisterous behavior. Each team played with one paddle. As soon as one player hit the ball, he placed the paddle on the table for the other to pick up and return the next shot. The men laughed and shouted and teased and taunted. Partners often ran into one another. They dropped paddles and missed shots. One evening Sam Roberts became so frustrated and angry that he slammed his arm and paddle onto the table, smashing his watch. Glass, fly wheels, and springs scattered across the surface and onto the floor. Everyone but Sam collapsed in laughter. He didn't know whether to laugh, cry, or rage. He stood in stunned silence while the others laughed on and on.

On Saturday nights the men and women set aside their businesses and families and enjoyed themselves. The men could be unguarded and playful—even a little adolescent. However, they were all serious people, good parents, accomplished professionals, and civic leaders. They lived with a sense of proportion and balanced their lives among family, friends, business, and community. They took pleasure in fun and friendship. They also depended on one another for personal support. It was a loyal, caring group. There was something deeper than just golf, poker, ping-pong, jokes, arguments, and parties.

These lifelong friendships, the other friends and acquaintances Ervin had across Delaware, and the relationships he sustained with his Columbus family and friends reveal an attitude toward people and a stance on life. He valued others and cared for them in remarkable ways. He might have been compensating still for the absences in his childhood and youth. But he was also making a full, rich life with and for his family, friends, and city.

"He made a difference in lots of lives," Louise said. "Ervie got inside a lot of people." And they cared deeply about him. The *Gazette* editorial said it well: "The shock that swept this community last spring when it was learned that Ervin F. Carlisle was going to die within a few months was a genuine sense of the place he held in the hearts of thousands of people. It was not only the shock that a man who enjoyed life to the brim was being taken away from us in the prime of life at forty-six. It also was the sudden knowledge that Erv Carlisle filled an even bigger place in Delaware than we had realized."

I knew most of the people who sent notes to my mother after my father died, but I did not know, nor did my mother, this man who described Ervin so movingly: "I cannot honestly say I knew your husband very well ma'am. That was my loss. After I left Delaware High, I saw him all too infrequently. But when I did, his demeanor and his attack on the world was always so cheerful and optimistic that the casual passerby like myself went away convinced that this wasn't such a bad old world after all. And believe me this is no small gift in a world that often seems so harsh and senseless and useless."

People are inclined to speak better than they might after a person has died and particularly after years have passed. Memory does select and invent. We are driven by a desire to create good pasts for ourselves and for the people we care about. Nevertheless, memory is sometimes based on realities—words and events reported in newspapers and documents, images captured on film, accounts in letters, and accurate recollections. Memory can be reliable—up to a point. It is not all wishful thinking. So I believe the words that Gene and Bob and others used about my father— warm, sincere, generous, firm, active, assertive, organized, brainy, honest, and principled.

But I had to be watchful, I knew, and also look for flaws and failures—search for the whole person, as well as the good and generous man I was discovering—avoid being sentimental and overly indulgent. I told myself, for example, that Ervin's determination to make things happen as he wanted, never to be satisfied with the status quo, to generate, as well as meet, challenges over and over might sometimes seem too insistent or aggressive. I also realized that he was a man of his times and acted once or twice in ways that now seem mistaken. I do not know if he believed in it, but he administered, for example, a wartime school board policy that prohibited hiring newly married women and that also required the automatic firing of any woman married during the school term.

I remember a few occasions when we passed by ideal picnic sites along the Scioto River because it was "too cloudy"—meaning there were too many blacks nearby. For all his professional commitment to nondiscrimination, Ervin could be modestly racist personally. Nevertheless, I realized that searching too intently for flaws could result in another kind of distortion. There are genuinely good people in the world—imperfect to be sure, but good.

It took several years and many roads back to find and portray the man—before and beyond father—whom I believe Ervin was. By searching and writing, I restored my father to myself and to my sister. "You gave me back my father," Linda said through tears, after she'd read my book about Ervin. She was just fourteen when he died. I found, as well, the boy, young man, husband, lover, educator, and community leader I'd sought. And I do love that *man*. I admire, and perhaps envy, his generosity and concern for others (for *all* others); his good will; his playfulness; his great capacity to enjoy life and love the people in it; his organizational and leadership talent; his friends, home, and community. I am impressed by the fullness of his world.

The South Side of Columbus shaped him. He knew loss and loneliness from his mother's death and his father's departure. Fortunately, Ervin was loved and cared for by his grandparents.

He grew up in an immigrant, working-class community. He took his first job at age nine. It was a life that challenged and tested him almost from birth. It could have been profoundly disabling. But Ervin somehow overcame all of that. He was a tough, determined, ambitious, and gifted boy and young man who transformed loss into gain and crisis into opportunity.

He overcame, but he never forgot. Nor did he ever reject that South Side life. He loved and was forever grateful to his grandparents. He enjoyed and valued his Döerflein and Stith relatives for the rest of his life. After graduation from high school, he established, with other former South High athletes, an organization to sustain friendships, stimulate college attendance, and further school standards throughout the city. He regarded the South Side as his part of town—as the home of his childhood and youth. It was his primal place.

Ervin transformed places of challenge and conflict into places of achievement and love. In Delaware, he made a place for himself, Winnie, and Linda and me like none he had known before. Delaware gave him a family, dear friends, positions of prominence, and a community. It enabled him to realize fully the self that emerged from South Side Columbus. At the height of Ervin's achievement and fulfillment, he died at age forty-six.

My mother lived to age ninety-one. I had her for most of my life. I knew her well. I lost my father, however, before I knew him as anyone but "Dad." Like so many young men—perhaps some of you who are reading me now—I had to come to terms with my father and with his early death. I have no other answer to those who've asked, "Where is your mother in all of this?"

My father, Ervie Carlisle, lives in my memory and imagination as a more complete person than I had known during his life and as a genuinely good man. Finding Ervin has given me a sense of rootedness—of a man in place. I came to that, however, through documents, newspapers, stories, interviews, and my own memories—not through direct contact with my father. He had died fifty years before my search began. Writing about the man has enabled me to understand and recover my own past and hometown—to the extent we ever understand our lives and ourselves.

11.

_____

208 West Lincoln Avenue

When I was three, we moved from 120 Campbell Street around the corner and eight houses down to 208 West Lincoln Avenue. Ten years later we moved farther down the street to 101. I lived on Lincoln for fifteen years until my sophomore year in college. The house at 208, however, was the first *place* I truly knew as home. I was not literally born in it, but it is emotionally and imaginatively my birthplace—my primal place.

When I think about that house and neighborhood, I am flooded with images that go on and on. I see the street itself where the old interurban trolley tracks ruptured the pavement and where I played pitch and catch with friends; I see the Rutherford and the Huntsberger houses on each side of 208, the Harter residence across the street, all the homes up and down Lincoln, then our wide front porch and the swing suspended from the ceiling on chains and springs, and then every room inside. I remember the smell of my parents' bedroom, the unheated upstairs sleeping porch where I sometimes slept even in winter under piles of feather quilts, the coal pile and furnace in the basement, the basketball hoop on the garage where I spent hours shooting baskets and making up games, the cherry tree behind the house, which I often climbed, the long slope of the backyard, and my father's Victory Garden at the bottom.

There is more to these images and memories, however, than simple reminiscence or nostalgia. The house, the backyard, the neighborhood, and then the woods in the center of a long and deep block shaped me significantly. We lived in a typical small-town, three-bedroom, one-bath, single-family residence separated only by the width of the driveways from the Rutherfords on the east and the Huntsbergers on the west.

The Rutherfords were old. She bought and sold antiques. He kept a large garden behind their house and killed chickens in his backyard. On those days, I watched him sharpen his hatchet as he pedaled a large circular grindstone, watched him chop off the chicken's head, and then stared as the headless chicken ran madly in circles and fell over. I last saw Mr. Rutherford lying in his upstairs bedroom, shirtless, red faced, sweating, after he'd suffered a heart attack. He died within a few days. I don't remember if we went to his funeral.

Mr. Huntsberger, "Dard" his wife called him, was a juvenile court probate officer; he'd been in law enforcement for a long time. He taught me how to play mumbly-peg and showed me where he tied fishing lures in his basement. I stayed with them a few times when my parents were out of own. After we'd moved down the street, Mr. Huntsberger shot himself in the upstairs bathroom. His granddaughter found his body.

Except for two doubles, one across the street and slightly to the west and the other two doors east, the houses on our block were all single family—some larger and some with wider lots like the brick Cherrington house. But by today's standards, all were modest. My own house looks very small as I drive by now—smaller certainly than it seemed when my family lived there.

The design and layout of a house—its geography —and its situation in the landscape or neighborhood reveal something about the way people live within it. The house expresses a certain form of life and also sustains that way of life. It enables and enacts it. When I lead you through and identify the locations and functions of rooms, I am in effect writing about the people who lived there. In that sense, houses tell stories about the lives of their inhabitants.

The house at 208 stood close to its two neighbors and close to the street. The walkway from sidewalk to front steps was only ten or twelve feet. The porch extended fully across the front. A porch swing hung across it at the far end. We must have also had chairs. I remember some vaguely—sitting in one while my father filed a ring off my swollen finger after he'd soaked my hand in ice water. But maybe I just want chairs because it suits my sense of home.

Not only was the house embedded physically in the neighborhood; it was also rooted there socially. The front steps and porch served as the transition between the public space of the street and the private space of home. The front porch, in effect, extended a Carlisle hand to the neighborhood and to Delaware. The hands of the neighborhood and town, in turn, reached toward our home by coming up the walk, ascending the front steps, and knocking or ringing the doorbell. In summer and during mild weather in spring and fall, my mother sat on the swing, snapping beans or shelling peas, reading, or talking with visitors and folks walking by. There wasn't a lot of foot or for that matter automobile traffic along Lincoln, so it wasn't as if people passed frequently and stopped to talk or that drivers regularly honked as they drove by. In my mind's eye, however, the porch established a relationship between private home space and the public.

The front porch also served as a meeting place between my family and the commercial world. Jimmy tossed a tightly rolled, boomerang-shaped *Delaware Gazette* onto the porch every Monday through Saturday; the milkman left glass bottles of milk at the top of the steps next to the railing post, and we returned the washed empties in the same place; the Omar bread man brought loaves and rolls to the front door; John left mail twice a day in the box next to the front door, and my parents placed outgoing mail in the box for him to take; Pocock's grocery, down the street to the east, delivered my mother's order, although Jack Pocock might have brought it to the side or back door; and there were others—less regular and sometimes less appreciated.

The front door opened directly into the living room next to the stairway to the second floor. The main area of the room, with a fireplace facing the front of the house, extended to the left. The kitchen was behind a swinging door straight ahead. The living room opened through an arch into the dining room. The family gathered there every day for dinner—my father seated at the head of the table, my mother at the foot, and my sister and I sat along each side. That arrangement—like the house as a whole—expressed a particular form of family life, distinct, say, from family meals at a round table or in front of a TV.

Behind the dining room, there was a small study where my father, once he left public school administration, started his "Personal Insurance" business. From the dining room, another swinging door opened into the kitchen where my mother prepared meals and cleaned up afterward and where she could look through two windows above the sink into a window of the Rutherford house

I remember Marjorie—a seventeen-year-old "colored" girl from the South End—assisting my mother with kitchen work and cleaning. She was a delight. She once told me I should have a pretty girlfriend because I was a handsome boy. Marjorie did not work long for us, however. She died. I don't remember how. Her younger sister, Pat, came to work for us soon after that. Later, when we'd moved down the street, Pat became an object of my friends' and my teenage sexual fantasies. We ogled and giggled but did nothing more. I did not understand at all the racial implications of this.

To the right of the sink, a door opened to the basement stairway—down three steps to the landing where the side door entered from the driveway and then on down to a basement, or cellar really, where my mother laundered clothes in one room and my father tended the coal furnace in the other. On Monday washdays, I would sometimes watch my mother dump dirty clothes into the 1930s washer, feed them through the wringer, place them in a basket, and then carry it up the stairs and outside to the clothes line in the backyard.

I had no furnace duties until we moved down the street. There,

I was charged with filling the automatic stoker from the coal bin—a task I performed rather sloppily, scattering pieces of coal on the floor near the furnace. At 208 my father used a wide shovel to feed coal through the furnace door and a long poker to stir the fire. I remember well that long red-hot poker—especially when my father used it to spear mice and rats. The squeals and stench are unforgettable.

The stairway to the second floor ascended along the east wall of the house. There was a landing two-thirds of the way up where the stairs turned left toward a hallway that bisected the house. I walked up and down these stairs for years without a thought of what it all meant. In fact, I lived for decades after that, deep into my adulthood, with no sense of the character and power of that place.

The hallway led directly to the small, three-piece bathroom at the end with tub, sink, and toilet—no shower, only a rubber hose with sprayer attached to the tub faucet. On the left, the hall opened to my parents' bedroom at the front of the house and right to the two back bedrooms. The second was my usual room with windows looking toward the Huntsberger house and the backyard. The first was my sister Linda's.

The first bedroom led to a sleeping porch that extended across the back of the house. My right knee still carries a scar from the day I broke the glass door on one of the lower bookshelves there. For a few years, I hung a small basketball hoop from the paneling and played with a small ball. Windows lined the back and side-walls and gave views into the backyards of our house and those of our neighbors. One day, guilt-ridden, I threw a small dish out of one window as far as I could into Mr. Rutherford's garden. I'd stolen it from a vacant house down the street when a group of us broke in through a back door.

When I look closely at 208, the mundane and conventional disappear, and the house becomes a place of meaning and power. The house embraced a family and separated it physically and psychologically from others—even if only by the width of the sidewalk, the driveways, and the street. The family shared living and dining room spaces—private and personal spaces for the most part. From time to time, guests were invited to share these rooms as well. My mother's special places were the kitchen and laundry room; my father's were the small study and the furnace room. Their spaces were largely private areas. The upstairs was our truly private family space. The design and function of the house told a story about mid-century gender relationships or identities, as well as about middle-class family life.

I am intrigued by the rhythm and flow of public and private associated with a house. At 208, space flowed naturally from the public areas of street and sidewalk, onto the front porch, into the house where living and dining rooms could be shared, to mother and father spaces, and then up the stairs to private family space. We reversed the flow from morning into the day as we descended the stairs and went about our daily business inside the house and then outside and down, up, or across the street.

The house centered and integrated family life in particular ways. It expressed and enabled that way of life, and it also shaped expectations and values. It helped form a personal and family culture, distinct, for example, from an urban apartment home and family life such as the one my friend Edward lived or the farm life my wife, Beth, lived growing up. Like other successful homes, 208 West Lincoln provided, in Witold Rybczynski's terms, "convenience, efficiency, leisure, ease, pleasure, domesticity, intimacy, and privacy." It provided, in short, comfort and a sense of order and security.

"The house we were born in," to reprise Bachelard, "has engraved within us the hierarchy of various functions of inhabiting. We are the diagram of the functions of inhabiting that particular house, and all other houses are but variations on a fundamental theme." A house is a personal, family, psychological, and social center for its inhabitants—the center of their lives. It becomes "the topography of our intimate being," according to Bachelard. "It is body and soul."

My home at 208 West Lincoln Avenue was my original corner of the world. It was and in some sense still is the center of my life. Once upon a time, I dwelled in it. Now it dwells deeply within me. Most of the houses I have lived in since have been variations of that primal, iconic place—mostly two-story, single-family homes in relatively small university towns.

I believe all this. But I am also aware of the ways the darker realities of life threaten the "protected intimacy" of a house. Mr. Huntsberger shoots himself in his upstairs bathroom. Mr. Rutherford suffers and dies in his upstairs bedroom. John Robinson, a high school boy showing off his handgun across the street, accidentally shoots himself fatally. I did not see the incident, but I crossed the street later that day to look at the bloodstain on the sidewalk. A troubled Dick Smith stands on our front porch with a stolen rifle and shoots bullets at the house across the street. No one died in that incident, but by firing that rifle from the space where public and private meet, he threatened our family dwelling and the well-being of the neighborhood. In one sense, these are everyday occurrences even in supposedly idyllic, small-town life. They are also existential events. And in these cases the house stood firm as a defense, a refuge—its "protected intimacy" intact.

Houses lined three sides of our rather large residential block—along Liberty Street, Lincoln Avenue, and Forest Avenue. Only a few houses had been built along Dent Street on the back of the block parallel to Lincoln. On the inside, behind the houses, there was a large wooded area my friends and I explored for years. The Woods.

I left the house, walked past the garage—grape vines covered its side—down the long, sloping backyard, between the Huntsberger hedge and the Rutherford fence and briar bushes, through my father's vegetable garden; crossed the old alleyway, passed along the edge of the Rutherford garden, turned slightly left near an old chicken coop, and then found the path into the woods—my boyhood sanctuary.

The path led first into a field overgrown with grass, weeds, scrub bushes, and small trees. It went on, crossing an old fence line, into a more heavily wooded area. On the left, a thicket of poison ivy vines had grown on the back fence of the Amy's yard and spread along the ground and up into trees—a dangerous spot, given my susceptibility to the itchy rash.

The path then crossed another fence line and opened into a stand of old apple trees. To the right, I could walk down into a briar patch behind the Cherringtons. They had terraced their deep backyard with limestone walls but let the last low piece go wild. It was impassable except along one side between the Cherrington and the Eishen yards. To the left of the orchard, I could walk through an unmown grassy area with just a few large trees to Dent Street. Straight ahead and beyond the orchard, I walked across another unkempt grassy area and down a bank into a depression behind the houses along Liberty.

I passed into my own world of exploration and imagination in the woods. My closest boyhood friend, David Smith, occasionally joined me. Sometimes, other friends came as well. We simply wandered, climbed trees, looked for small animals and birds, tried to avoid the poison ivy, or played war games or cowboys and Indians. I also went to the woods by myself. And then its character became clear—as I think of it now. It was the geographical center of the block, and it was also an escape and refuge from my family and neighborhood—a place of solitude and distance. The woods suggested the pastoral, the rural, even something of the primitive—at least as I imagined it. I could escape to a protected place yet imagine something of the wild.

David and I had another boyhood escape. Several times each summer, we followed the Delaware Run from Elizabeth Street, past

Sanborn Hall (the Ohio Wesleyan music building), through Blue Limestone Park, past the old quarry, over a high railroad embankment, along the bottom of the Jane Case Hospital grounds, and then on out of town. We found dead animals lying along the Run—a large, decaying dog one day. We threw stones at water bugs. We captured crawdads. We could never catch a frog. Beyond the embankment, the Run passed through open, undeveloped, and often overgrown land. It seemed rural to us—a bit of the wild, perhaps. Once in a while, we'd go as far as Houk Road, almost a mile past the hospital, and walk back into town along State Route 37, Central Avenue. At the time, it seemed a long way out. Now, the town, four times the size it was then, has expanded well beyond Houk Road.

These explorations into nature, as they seemed at the time—like the house and the neighborhood—helped shape my sense of space. They formed my expectations and desire for space. They helped form my personal culture. They provided separation, freedom, and safety—a place of solitude and a place for contemplation insofar as I was capable of that. Later in life, when I could, I gravitated toward country houses and space—as I did with Three Meadow Mountain in Clover Hollow.

The woods also had a dark side—or became the scene of my dark side. One day a friend and I led Mr. Rutherford, our next door neighbor, into the woods to show him where several kittens had been buried alive—by some boys, we asserted—and maybe beaten before they were. We helped him dig them up—I think—and there my memory stops. It's all so vague. I can't help but wonder if I did it. How would I know otherwise about the kittens?

The woods are gone now. Homeowners cleared it away. Houses were built decades ago all along Dent Street—later renamed Elmwood Drive. Backyards and gardens now extend from one side of the block to another. When I drive along Elmwood, I can see all the way through to the backs of the houses on Lincoln. The woods are truly gone. But still I wander in memory and daydream through my sanctuary.

I left the woods the same way I'd come in—following the path through the first overgrown area, passing the chicken coop and the gardens, crossing the old alley, and then walking up the yard toward the house.

In winter in that backyard, my father and I would spray the snow cover with a garden hose to make an icy, downhill sled run. It started next to the garage, curved behind it, and turned left—where we'd banked it and built a low wall of snow and ice—to run straight down the slope, across the garden, to the alley. One winter, we built an arch of large snowballs to sled through. I would lie face down on my father's back, and flat on his stomach, he would guide us down the hill and under the arch. "Keep your head down, Freddie." Once I did not. I looked up, smashed full face into the arch, flew off the sled, and ended up on my back amidst a pile of broken snowballs. My father had glided on through. We did not rebuild that winter.

David Smith planned to spend the night—something we often did at his or my house. We had pitched a tent in the backyard just below the windows of my father's office. David, Susie Semans, and I were sitting in the tent talking. One of us must have said something obscene and then laughed. "Freddie, would you come into the house, please," my father said. I went in. "David will not be staying over. If you and he can't behave like gentlemen around a girl, you shouldn't be together." I learned other lessons from that backyard, as well as from the house.

Friends and I—I don't remember which ones—decided to hold a small backyard carnival at 208. We offered prizes for games we'd devised—I don't remember them either. I do remember, however, that when Mary Keller came with friends, she started crying when she saw some of the prizes. No wonder. I had taken them from her house a few days before. Lying, dishonesty, and stealing were the worst offenses in our household and brought serious punishment—this time a rare whipping by my father. But even worse, my father made me return Mary's toys to her house and apologize face to face to her parents and to her.

He made me do the same thing after I had shot two younger Campbell Street boys, Phil Hoffman and John Jeisel, with my BB gun as they ran away and tried to hide behind a garage. I walked over to Campbell Street. I knocked on each door. The boys' fathers answered. I told them why I was there. They were not visibly angry—but hardly warm and friendly. They accepted my apologies. I walked back home appropriately shamed and humiliated.

I also spent many satisfying hours by myself in the driveway, alongside the backyard, shooting baskets at the board and hoop mounted on the front of the garage. Doug Dittrick, a friend and neighbor from around the corner on Forest, joined me once in a while. But mostly I played by myself—making up games and tournaments using names of the famous college teams of the day.

The backyard extended the interior of the house to the outside. It was family space and also a place we sometimes invited friends and guests to enter, just as we invited them inside 208. It was the site for leisure and also for lessons—minor ones perhaps but lessons I remember well. The backyard expressed and further enabled my family's way of living—its form of life.

———

When I write about hometown and my original corner of the world, I am relying significantly on my own memories and somewhat less on records, documents, and the recollections of others. I understand the limitations—even the unreliability—of memory and also the limitations of historical records and written histories. Memory follows the line of time backward. It is, we would like to think, based on realities. But it also transforms those realities through ignorance, exclusion, interpretation, and invention. It invests feeling and interest in the past and thereby gives it energy and meaning. It is driven by desire.

History, on the other hand, follows the line of time forward and informs memory. It tries to tell a reasonable and likely story about the past—or so we believe. But history itself can be incomplete, selective, and occasionally unreliable. It is also based on desire— the desire to make the past orderly and intelligible.

Even if memory and history are imperfect, they are all we have of the past. So we must do our best—taking an ironic perspective that on the one hand values and relies on memory and feeling and draws on historical material, but on other hand, through thoughtful and informed analysis, reveals our distortions and inventions regarding the past. We seek to make life understandable, but we also must recognize what we're doing.

I am also aware that I'm writing about myself as a small-town boy—a character, in a sense, that my adult self has created from my own limited and selected recollections, and to some extent from ideas and texts as well. Many of you might not share this specific kind of experience; nor would you represent yourselves as the same youthful character I've invented. Nevertheless, I believe our original corners of the world and other primary places in our lives have shaped us all in similar ways. We share equivalent experiences. I am writing to encourage you to recognize the power of place in your life.

12.

_____

The Delaware City Schools

North Elementary

North School was one of four elementary schools in Delaware—each named for its part of town—North, East, West, and South. North was located on Washington Street in an older area of the city. Kids walked along quiet residential streets going to school and returning home. Not many rode bikes—even though bicycles were our primary way of moving about town. There were no buses, and parents did not typically drive their children to school.

North provided a psychological, educational, and cultural center for the families in that end of town. It focused and ordered an important part of my family's life. And now it is a site of memory for me. I remember every classroom, teacher, and the unusual geography of the school—the layout being my main reason for leading us around the interior, from classroom to classroom.

Built in 1869, North was old even in the 1940s. It was a red brick building with large windows and a tall bell tower above the front double-door entrance. It was constructed as a conventional rectangular building, but within it was laid out as an octagon with classrooms opening off a large central lobby or rotunda. Myrtle Bell, who seemed almost as old as the school itself, taught fourth grade in the first room to the right of the school entrance. We sat on bench seats at fixed desks, each with an inkwell on top and a shelf underneath for pencils, papers, and books—a truly old-fashioned space.

A youthful Mrs. Warren taught second grade in the next room. There we sat on small chairs at low tables. Her room opened at the far end to the playground and softball field. I remember nothing about second grade except for Mrs. Warren's face, the layout of the room, and my father, the acting superintendent for Delaware, sitting on one of those tiny chairs eating a bowl of soup we'd served. I cannot name the occasion.

Double doors opened next into Mrs. Easton's third grade, where we reviewed multiplication tables over and over, sitting at the standard fixed desks. In her room, we organized a classroom grocery to help learn arithmetic. She pronounced the word "chocolate" in such a delicious way that I've never forgotten.

Principal Tippett (a formidable man with a big mole on his left cheek) occupied the room opposite the school entrance. When I was in fourth grade, he paddled me for grabbing and shaking a newly planted tree in front of the school. It was my second spanking at North. My first grade teacher, Mrs. Klamforth, gave me the first—a hard slapping on my bare leg where she'd pulled up my shorts.

To the left of Harold Tippett's office, Nancy White taught fifth grade. She was a very pretty woman. Then came a long hallway back to first grade, the only classroom not on the rotunda. The first-graders had a separate, outside entrance at the end of that hall. Edith Swartz's sixth-grade room was the last one facing the rotunda.

The architecture of North Elementary unified the school in a particular way. Except for first grade, all students entered and departed through a single front entrance. The classrooms opened to a large, central rotunda where students and teachers mixed going to and coming from class. With first grade at the end of a long hall, the geography protected the youngest students from the older ones and then joined all the rest of us together. The design expressed and enacted the common educational and social purposes of the school. It centered and unified the school in a real, if indirect and tacit, way. And I believe—even if I cannot quite say how—that North differed in its effects from schools constructed with rooms opening off long hallways and with more than one school entrance—like West, the school my friends David Smith and Lloyd Gardner attended.

Outside, North's playground bounded the school on three sides. At the back, kids played on swings, teeter-totters, a merry-go-round, and a Jacob's ladder. The area was open on the north side. We used the south side for a softball field—narrow and gravelly as it was, except for a worn, grassy area in the outfield. On that field, as well as in Delaware generally, I began learning about race.

In the mid-twentieth century, Delaware was sharply segregated residentially and socially. The blacks lived almost entirely in the South End. Only one or two black families resided elsewhere. Delaware still required them to sit in the balcony of the Strand movie theatre. It excluded them from swimming in the public pool at the fairgrounds. Even when blacks were finally admitted, the pool set certain days when they could come. On those days, my Auntie Toots—Lillian Thomson—refused to go to the pool.

The elementary schools played softball against one another with teams of mainly fifth- and sixth-grade boys. For white North Enders like me, South School seemed a foreign and frightening place. It was a two-story brick building as old as North with a playground as hard and bare as ours. That's where we played the games.

I knew very few black kids other than the girls who worked for my mother and a group of boys who hung around the high school football field during practice. When my father was helping coach the team one year, he sometimes took me along to practice, and there I played touch football with boys from the South End. Except for that small, friendly football group, I knew black kids only as intimidating softball players—loud, bold, strong, and threatening. We were playing on their ground—their place. They made sure we knew that. Space and place not only are sites of identity but assertions of it as well.

I shared the prejudices of the day and saw with those biased eyes. At the integrated junior and senior high school, a few years later, young blacks like Cy Fleming and Jack Miller intimidated boys like me to establish their space—their identities—in that mostly white place. What else did they have?

Those softball experiences, neighborhood segregation, and the white prejudice of my social class, clearly shaped my youthful attitudes. We usually lost, by the way, when we played South—9–8 and 12–4 in the sixth grade. That same year we lost big to East, 32–13. The year before, however, on North's gravelly field at home, we actually beat South—I think. But I might have invented the memory; I've not been able to verify the score.

I also began learning about social and economic class differences at North. Parts of the North End were more privileged and afflu-ent than other parts of town—at least in my neighborhood and among my parents' friends and acquaintances. Kids from my neighborhood went to North. The school also drew kids from social and economic classes different from mine. I remember par-ticularly two boys—John Boyer and Gerald Asbury. For a short time, John and I became friends. Occasionally, after school on Wednesdays, we'd walk to the high school to see the afternoon movies shown there. As acting superintendent and with his office there, my father arranged for us to see the shows. John might have come to my house a few times. Once or twice, I went to his on Blymer Street—a lower-income, working-class area. John wasn't bright, but he was a nice kid whom I liked. The friendship trou-bled my mother, however. I shouldn't have a friend, I guess, who lived down on Blymer near the river. He came from the wrong class of people.

I don't remember Gerald well—except for his bright eyes and freckled face and especially for his edgy, aggressive question one day at North—"Did the jizz fly last night?" Rumors and sexual innuendos had been spreading about me, a girl, and maybe another boy. They so disturbed the school principal, Mrs. Mathews, that she interviewed me, and perhaps others, to find out what, in fact, was going on. I don't recall the outcome, but it all ended with her intervention. Gerald, from a working-class family, introduced me to another style—a more sexualized and aggressive behavior.

Like other places, North wasn't simply an occasion or a neutral place where people interacted and events occurred. The school not only expressed, it also enacted certain forms of life and influenced its students significantly. We learned to read, write, and compute from our teachers, the authorities, and we behaved by rules enforced by administrators Mr. Tippett and Mrs. Mathews. We felt safe and secure in this small community where we presumably shared common educational and social purposes with other students and teachers. We were all white—if from somewhat different social environments and classes. We learned new things about people from those differences. Nevertheless, we mixed reasonably well. Occasional playground arguments or fights didn't threaten the underlying order. Teachers or the principal ended them quickly. Both the location and the architecture of the school reinforced a particular form of life. The place, also, told a story about life in Delaware.

Eventually, after a new North was opened, when I was in high school, old North became the city's Youth Recreation Center. Later still, it became the school district's administrative offices. In both later guises—youth center and district office—North School was an altogether different place, expressing other forms of Delaware life.

Frank B. Willis High School

I went from North Elementary to Frank B. Willis High School—a seventh through twelfth grade building—when I was eleven years old. I spent six years there and graduated at age seventeen. Other than my home, Willis was the center of my life for those years and a place where I changed in virtually every way—physically, emotionally, psychologically, and intellectually. When I think about that time, images and memories tumble through my mind almost uncontrollably.

I remember teachers, classes, and classmates: the football coach's health class—Ray Overturf dropping his trousers to show a wartime scar; Mrs. Zimmerman falling asleep at her desk in English class; Mrs. Evan's Latin classes where, to my surprise, I was one of the few to translate Caesar rather than rely on a "pony," or cheat sheet, and where a few times, sitting next to a side wall blackboard, I dragged my fingernails down the board and electrified the entire class; Sherman Moist's social science classes where I studied but did not really understand social issues and conflicts; Raymond Felts's math classes where we learned algebra, geometry, and trig, and where we also listened to the World Series while we solved problems. And then the supreme Dorothy Whitted's demanding but wonderful junior and senior English classes where we read great books—the Russians, as well as the Brits and Americans—as she tried to prepare us for college. When I was a sophomore, I would occasionally look into her classroom during an exam and feel totally intimidated.

When I think about athletics, games, road trips, teammates, individual opponents, and coaches flash through my mind. I scored twenty-eight points in the first basketball game of the season, struck out twelve Grandview players in a five-inning summer game, never threw another inside pitch to George Lynn after he hit a home run across a street onto a front porch. I was captain and MVP on a losing basketball team, 3 and 14, in my senior year. That same year, we won more baseball games than basketball—how could we not? A Columbus newspaper selected me as one of two pitchers for the All-Central Buckeye League team.

And I recall high school activities and social life. Girlfriends, of course. My first at Willis—a farm girl with very strict parents; my second, a seventeen-year-old senior when I was a fourteen-year-old sophomore; then, my great high school love, a blond cheerleader who gave me emotional highs and lows that still amaze me. And then less emotional activities like raising money for the Junior-Senior Prom by selling Christmas cards, running the noon-time popcorn stand outside the doors of the gym where kids gathered after lunch, and managing the hot dog concession next to the football stadium.

But tempting as nostalgic reminiscence about high school might be for many of us, I am writing, instead, about Willis High School as a complex *place* in all the meanings of that term. Willis was a constructed, psychological, social, cultural, and imagined space as well as a location. It was yet another place that profoundly shaped the person I am.

Opened in 1932 as a high school and later a middle school, Willis is a three-story, brick structure originally built as a quadrangle around an open courtyard. The auditorium—the space for study hall, school assemblies, and music and drama performances—stood on the west side of the main building and faced William Street. The gymnasium, also facing William Street, stood on the east side and completed the architectural symmetry of the front facade. A one-story wood and metal shop wing extended perpendicularly from the rear of the school.

The architecture was typical of 1930s school design—long hallways lined with student lockers and general classrooms opening into hallways—each room with chairs or desks organized in rows, the teacher at the head of the class. The structure also provided specialized rooms and spaces for art, music, drama, recreation, performance, wood and metal shop, home economics, and agriculture. These spaces were less rigidly organized, although the art room was laid out with individual art desks in rows—in no sense a studio arrangement.

Image 9. Frank B. Willis School, ca. 2012. Image by author.

The design expressed and enabled a certain form of educational and social life. That is, the exterior symmetries and interior order expressed a formal, teacher-centered style of education. Willis High School, in a certain sense, could have been almost anywhere—its formalities were so common. My classmates and I, no doubt, resembled high school students in many towns and cities. We were learning typical subjects taught by teachers with generic college educations. We were experiencing the awakenings, confusions, fears, and tortures of adolescents everywhere.

But in a more reflective sense, Willis High was unique—existing at a specific time, in a particular town, with my schoolmates and me in attendance—and with its unique sounds and smells, marks on classroom tables and desks, notices posted on bulletin boards, "dirty" graffiti scrawled in the toilets of the old building, mice eating through lunch bags stored in student lockers, our own dust and dirt, and our own particular events, games, and performances—all that made it *this* place and no other.

Willis also played a special role in the spatial and social geography of Delaware. The school is a block and a half from the main downtown intersection of Sandusky and William Streets. Occupying half of a city block, the school was a dominant physical, educational, cultural, and symbolic presence near the center of town. It was the only junior and senior high school in the city and the township. As such, it gathered together the entire town and surrounding rural areas and brought classes, races, and lifestyles together at the city's education center. It expressed, thereby, a common purpose—as the octagonal, interior design of North School did for grades one through six. It was a place that enabled—that even forced—encounters, relationships, conflicts, and friendships to occur. At Willis, I met people and experienced events that my protected and privileged North End life had not prepared me for.

Six years of classes made some intellectual difference in my life. I read a lot of books. Like high school students across the country using common textbooks, I read *Julius Caesar* in tenth grade, *Macbeth* in eleventh, and probably *Hamlet* as a senior. I plodded through *Silas Marner* one year, another common reading. But not every high school student took on *Crime and Punishment* or *War and Peace* as Miss Whitted required. I wrote themes. I studied Latin. I learned elementary math—algebra, the geometries, and trigonometry (calculus had not yet descended into high school). I knew something about government and society. I'd become a student leader. I was more or less ready for college.

Those years also made an enormous difference in my emotional and physical development. I matured physically. I awakened sexually. I suffered through a tortured adolescence. Willis provided the stage—the place—for these awakenings, as it did in some way for every student in my class—and I'm assuming, as high school did for adolescent students everywhere.

Beyond these changes, I am also trying to understand the impact of the school on the community—as the only fully integrated place in the city. And in that light, I think about race and class—about the social and moral role Willis played in Delaware—in short, about the school's social and moral geography. Willis expressed a value and way of life the city did not really share and the school realized only partially.

I might be more conscious and troubled about race than many, but race remains a serious problem for us all. "It will trouble us a thousand years," Robert Frost's narrator says about the problem in America of realizing Jefferson's principle "That all men are created free and equal." W. E. B. Du Bois observed in 1903, "The problem of the twentieth century is the problem of the color line,—the relation of the darker to the lighter races of men." He might have been writing about the twenty-first century as well. Our country has obviously not dealt well with race; nor has it addressed its class differences.

Racial difference and prejudice ran deep in Delaware. In addition to the restrictions at the Strand Theatre and the county swimming pool, residential segregation was virtually absolute. Community leaders even performed in local minstrel shows—and in blackface—as late as 1950. The pastor of the William Street Methodist Church, Sheridan Bell, and a prominent local attorney, Clyde Lewis, were end men.

I had encountered blacks at home (the girls who worked for my mother), on elementary school softball fields, and during high school football practice at Community Field. I liked Marjorie and Patty—but then we were the "masters" in that situation. I was anxious, even a little fearful, during games at South School. I enjoyed touch football with the black kids from the South End. I felt intimidated by Cy Fleming's size and strength. I had known both ease and anxiety where blacks were concerned. That continued for several years.

Billy Ufferman and I sat on a cement step in front of West School, where the Youth Center occupied the ground level. Jack Miller, a black eighth grade classmate, stood over us and challenged our racist beliefs. "You guys think we all carry knives. You think we all smell—that we don't wash or wear underwear. You think we're mean and always ready to fight. You believe we're not as smart as you are. You people believe that colored people are not as good as you." Jack talked for some time—accusing, resentful, and a little angry. Billy and I didn't know what to say or do. What *was* there to say? He was probably right.

At my first school-wide assembly at Willis, the student body president, looking sharp in a V-neck sweater and slacks, appeared from behind the stage curtain to call the students to order. It was Briss Craig (Chester), one of the young blacks I had played touch football with. I was surprised, I think, but not troubled. His election was historic for the town. It was, as the *Delaware Gazette* reported, "noteworthy for the election of a colored youth to the highest office in the school for the first time in history." At that moment, Chester Craig was in charge.

In junior high, when assertive black males were establishing their place, I often felt anxious and intimidated. Cy Fleming seemed even bigger than his six feet and two hundred pounds when he'd grab me from behind in a bear hug. One day when I was sitting on the curb along the running track, Ray Broadnax stood over me, grabbed the back of my neck, and pushed my face down almost to the cinders. I don't remember why, but he left me in tears. After a few years, once lines were drawn, identities clear, things settled down—at least as I experienced them.

Blacks, whites, city kids, and country kids sat together in integrated classes where we were intellectual and social equals. But at lunch in the cafeteria, students typically segregated themselves—as we did in the gym after lunch where there was music and dancing. We were there together, but I do not remember whites and blacks dancing with one another. Willis was an integrated, central high school where kids of different races, backgrounds, and social and economic classes joined together in school activities, generally respected one another, and often were school friends. But those relationships—especially across the racial divide—typically ended at the school door. We left the building for our respective parts of town and entered the race and class structures of Delaware.

Athletics realized most fully, for me, the common purpose of Willis High School. Jack Miller and I became basketball and base-ball teammate friends, as did Joe Reed after he came to Willis as a junior. We dressed, undressed, and showered together in a damp, poorly ventilated basement locker room that stank of stale sweat and the toilet. We talked, kidded, and enjoyed ourselves as equals and friends. We practiced, played games, and won and lost as a team. Strangely enough, when we were seniors and won only three basketball games, Joe Reed and I were named to All-Central Buckeye League honorary teams.

"The football team got along really well," Bill Cooperider, a white classmate, explains, "because we were a team and that was told to us every day. Two black players started on offense and one on defense. My personal feeling is that Joe Reed helped pull us all together. Dick Burke, also black, was another guy that played a big role in the blacks and whites getting along well together."

We felt comfortable on the basketball court, the baseball and football fields, and in the locker rooms—as we did generally in the school. But in another part of town, away from that central place, we might feel out of place. The parents of a white reserve basketball player hosted the team for a meal at their North End home. In that house and that part of town, Jack Miller, dressed neatly in a coat and tie, looked so tense and uneasy—as uneasy as I would have in his end of town. At that moment, he clearly felt out of place—*place-less*—and in a real sense, out of self—in a world he neither knew nor understood—in a place that was not his or him.

Several old Delhi yearbooks from Willis remind me of other

ways blacks and whites joined together at school. Blacks constituted no more than 10 percent of the students, but for me, their
presence was somehow greater than their numbers. They were
involved in Future Teachers of America, National Forensics, thes-
pians, and choir. Two of the five cheerleaders were black—sisters
Phyliss and Eljean Green. The academic achievements of two
black students exceeded everyone else in the class, proving their
academic integrity and equality. Shirley Walden and Phyliss Green
were named to the National Honor Society as juniors and were
ranked among the top three in our graduating class. Shirley and I
were co-valedictorians. I spoke at the commencement because—I
would like to think—I was president of the student body and not
because I was white and male. Shirley—a gifted young pianist—
played for the processional and accompanied the cornet trio.

Frank B. Willis High School enacted a value and a way of life—an
integrated form of life—more completely, so far as I know, than any
other place in Delaware. Nevertheless, a sharp racial divide persisted in the city and obscured—or at least softened—real social
and economic class differences. The students from the North, East,
and West Sides of town were all white, and that made class differences less apparent as well. School activities and athletics, moreover, evened out the differences in ways similar to how black and
white got along. And in those teenage years, boy-girl relationships
were so charged sexually and psychologically that class differences
disappeared in the flames of adolescent love.

Two of my serious high school girlfriends came from backgrounds and expectations quite different from mine, but at the time
that made no difference. My first girlfriend lived on a farm north of
town. Her parents did not allow dating and even forbade her from
attending school dances and football and basketball games.

She was my at-school girlfriend only, except for a few months
when she stayed with a relative at the other end of the alley, a block
from my house. We met there a few times, but we didn't go out. She

soon returned to the farm, later moved to Columbus, and within a year or so she was married at age sixteen—unimaginable in my and other middle-class families. She and I had nothing in common except that, as one of my friends put it, she was "so enticing."

By then, I'd moved on to the lovely blond cheerleader who gave me fits for more than two years. Ours was a teenage romance with highs that thrilled me and lows that sent me to my sick bed. The cheerleader came from a working-class family and lived on the West Side, near the high school athletic fields and the South End. Her parents were divorced. She lived with her mother and elderly stepfather and occasionally with her sister in the North End. Her father drove a soft drink delivery truck. I do not remember her mother's work. Neither had attended college. Ruth tried Ohio Wesleyan for a semester, but she either flunked out or ran out of money. She wasn't especially bright, but oh so attractive and alluring and oh so aware of that.

As a place that brought kids from different worlds together, Willis became the stage for intense pubescent and sexually stimulating encounters that sometimes crossed class, but not racial differences, insofar as I knew—except in white boys' fantasies. There was obviously no future for Ruth and me beyond high school simply because we were so unlike. My parents and teachers knew that. I finally did. I was learning about class, in those relationships, without really understanding what had happened other than it was "natural" for high school romances to end.

Some years ago, I talked with a group of my former classmates—all of whom had stayed in Delaware and were people I'd not known well. Only then did I understand fully—and from their perspective—the class distinctions at Willis and in the town. I had gone on to a life beyond Delaware, and for decades I hadn't thought seriously about high school and hometown. My classmates, however, helped me return to my past and hometown and begin to understand my life there.

Two white classmates particularly—Judy from the East Side and Bill from the South End—spoke most clearly about their feelings and about class differences. At school, Bill felt "shunned," sometimes. "I lived in the South End on Harrison. I felt shunned and put down because of where I'd come from." Bill and I knew one another at school—there were only a hundred students in our class—but we weren't friends. However, the summer following graduation, we worked together on the construction of a new house in the North End. We got along well, as I did with the carpenters and other builders. In a way, I'd entered Bill's and the carpenters' world that summer and later during college summers when I worked construction. I crossed class lines, but only temporarily. In doing so, I sensed the distinctions, even if I could not articulate them.

Judy did not know she was poor until she crossed the river and came to Willis where she encountered kids from other and more affluent parts of town—kids, I suppose, like me. She said, "We were poor [on the East Side] but so was everyone else." From childhood on, Judy and the others worked cleaning houses and delivering newspapers, in her case. They expressed some resentment toward the elite of Delaware—the new elite, especially, such as an "outsider" city manager. The transformation of Delaware into a city they no longer knew saddened and troubled them. But that is another part of my story and not about our years at Willis.

I think of Willis High School—and North Elementary—as places of complexity and power that conditioned the way I live and shaped who I am.

Willis was a central high school *located* near the center of the city. Because of its location and character, it expressed a belief in orderly and unifying education and an integrated society—at least in public—even if it did not fully realize either belief.

It was a place of considerable *psychological* force: We were experiencing the difficulties and anxieties of learning, as well as its satisfactions. We were enduring the turbulence of adolescence. We were encountering people from other parts of town, another race, and other social classes and figuring out how to get along.

Socially, it was the outward manifestation of the inward dramas. We had entered a world of racial and class diversity. This new social world disrupted, on the one hand, the social structures we'd learned at home, in neighborhoods, and elementary schools. But on the other, it sustained those structures at deep behavioral levels. We shared classrooms, but we separated into social and racial groups elsewhere—except for athletics and other school activities. We were surrounded by the social structures of Delaware.

Willis was *cultural* space—a place of conventional school and small-town forms of life. The symmetry of the building, the orderliness of the classrooms, and the teacher-centered classes represented a style of education that emphasized the giving of knowledge, learning what you're taught or told, and the assurance of answers; if the answers weren't in the back of the book, the teacher knew them. Willis represented a relatively orderly and simple way of life. It also recognized the diversity of its population—in its tracked curriculum and in the racial and social differences that persisted within the school and guided student behavior. Willis enacted a culture of simplicity and clarity on the surface, but at the same time it sustained a culture of permanent racial and social difference and tension.

13.

Downtown Delaware

When I was fourteen, I worked part-time at Wilson's, "C. J.'s of course"—the elite downtown clothing store in Delaware where my father and his friends bought their clothes. I dusted boxes, restocked shelves, occasionally pressed garments on the steam press, swept the floor at the end of the day, and on one or two Saturday mornings, as I recall, cleaned up vomit left at the entrance by a late-night drunk. I waited on customers a few times, but only if they were acquaintances or friends of my family.

I lasted only a few months. Tom Wilson, the owner's son, fired me. "We won't need you any longer, Freddie. In fact, we won't need you after today. Turn in your hours and pick up your check next week." I'd made mistakes—like leaving dust and debris on the floor after I'd presumably swept it clean. And they probably didn't need a fourteen-year-old boy anyway. But the steam press incident doubtless made the difference. I forgot to turn it off. As a result, it blew up with an earsplitting whoosh and then loud hissing to the alarm of everyone in the store.

I mention Wilson's, the premier men and women's clothing store, because it represents what downtown Delaware was like. It was the business, commercial, professional, and in some sense, the social center of the city. In just one block of Sandusky Street—main street Delaware—people could pay utility bills at the Delaware Gas and the Southern Ohio Electric companies (and their telephone bills at the Northern Ohio Telephone a half block off Sandusky). They could buy auto accessories at Western Auto. They could purchase groceries at the A&P or at Hepner's Bakery, leave clothes at the Economy Laundry and Dry Cleaners, bank at two savings and loans, and then shop at Sears and Roebuck, the Boston Store, Blackburn's Clothing, Delaware Hardware, LeRoy's Jewelry, Bob Burns Shoes, Rinker's Drug Store, Willis Paint and Paper, Lyons Electric Appliances.

In the two Sandusky Street blocks, between Central Avenue and William Street, there were three drugstores (and another, the Winter Street Drugstore, around the corner on Winter), five food stores, eight restaurants (with the famous Buns, its sign arching over the street, just a half block west on Winter Street), four shoe stores, five clothing stores, three jewelers, three department stores, three hardware shops, two five-and-dimes, two banks, three savings and loans, two paint stores, one appliance shop, a furniture store, three auto accessory places, and numerous insurance, law, and real-estate offices, as well as attorneys, doctors, and dentists.

Delawareans could buy Oldsmobiles, Cadillacs, Fords, Mercurys, and Lincolns no more than a block away from the main street, and they could have cars repaired at those agencies or at Tilton's Garage right on Sandusky. I have not in fact included everything in those two blocks, or identified places along Winter and William Streets, or on South Sandusky where there were also businesses, stores, bars, and restaurants. The complete list would be very long.

At that time, downtown Delaware was effectively contained on every side, so that it could not expand beyond its historic center. It was bounded on the south by Ohio Wesleyan University, on the east by the Olentangy River, on the north by county buildings, the library, a church, and residences, and on the west by churches, the high school, and elegant nineteenth-century brick homes along Winter Street. It was the geographic, as well as the commercial and social center of the city. Years later, as the city grew and the older downtown was eroding, business and commerce expanded by developing open land south of the city toward Columbus. Over time, virtually all of the downtown businesses closed or relocated. Sixty years later, Sandusky differs dramatically from the street I knew.

Delaware was then a small community of 9,500 residents.
People knew one another. Virtually all of the businesses and professional offices were owned or managed by residents who cared
about the community and were invested in its development and
well-being. Only a few people worked in Columbus or anywhere
else out of town. Delaware was the county seat—with the courthouse, sheriff's office, the jail, and the locus of county government.
It was also the commercial and cultural center of the entire county.
It was the place to go.

On Saturday night—and then later, on Fridays—everyone went
downtown. "It was just policy," Gene Peebles, a long-time resident,
said. There was a "strong sense of community in those days, a real
camaraderie." The stores stayed open. Townspeople came to shop,
go to the movies, or simply walk along Sandusky, greeting friends
and acquaintances. Farmers and their families also came to town
to shop or simply park along Sandusky, sit in their cars, and watch
the passing scene. It seemed odd to me—all these country people
wearing country clothes, bib overalls, jeans, plain dresses, sitting
in cars along the curb. I did not understand then what it meant—
that Delaware was the county seat, the center in every respect, nor
did I know the kind of close-knit community it was. I was a kid,
living inside it, and unaware.

My home, North Elementary, and Willis High School centered
and organized critical parts of my life as the schools also did for
the city. The downtown centered the city's commercial, professional, and social life, as well as an important part of mine. It
helped organize small-town and rural forms of life. It *expressed*
it through the people, stores, and offices located there. It *enabled*
people to satisfy virtually all of their commercial, business, and
professional needs and also to join together in a community. It
enacted that form of life simply because people had to go there to
conduct business, to shop, and to see others. There was nowhere
else to go—except to Columbus, an hour away—which my

family did occasionally, mainly to visit my father's relatives but also, on special occasions, to shop or see a medical specialist. There were professional offices, neighborhood groceries, restaurants, and businesses elsewhere in Delaware like Pocock's grocery (later Allen's) and Eckleberry's Confectionary in my neighborhood. Not absolutely everything happened downtown, but a whole lot did. It was the actual and symbolic center of Delaware, Ohio.

At the same time, like Willis, downtown Delaware did not fully realize its small-town ideal. The social structures of difference and hierarchy persisted. One of my classmates told the story of trying to buy a suit at Wilson's. Howard was dressed in clean work clothes, but even so, "They wouldn't wait on me. They just wouldn't. So I went across the street to Anderson's and bought a nice suit. I never went back [to Wilson's]." Judy told me about a farmer who wanted to buy a Cadillac. He was wearing clean farm clothes when he went to the John Mathew's agency on West William Street. "How much is that car?" he asked. Instead of answering, the salesman said, "You can't afford it." So the man went to Keefers, bought his Cadillac, drove to Mathews to fill his gas tank, honked, and drove away.

Downtown did not bridge the racial divide well either. Some folks from the black community shopped at a downtown grocery, the U.S. Store, but not at many other places. Even now, some of the older blacks will not go downtown. In general, class and racial prejudice persisted in Delaware, as did residential and social segregation. The Delaware ideal failed, in part at least, its working-class and black citizens. Nevertheless, many of us lived as if Delaware were a centered and stable community—and in a relative sense it was.

My Delaware provided a relatively centered, simple, stable, and secure life. I grew up in the town and internalized its forms of life. They became the model for how things are and would be—or at least should be. Those forms not only shaped identities; they also shaped consciousness. I perceived, felt, interpreted, and thought within them. The effect was powerful and deep. My parents and my home—my primal place—gave me a secure, stable, and loving ground. The schools and the town confirmed that form of life in more public ways.

"We owned the whole town," my friend Lloyd says. "We could walk or ride bikes anywhere." And we did—although I didn't venture into the South End walking or riding, or even cross the river to the East Side. My own uncertainties and anxieties controlled that. It felt like we could go anywhere safely. "When I was a kid growing up," my friend Jack says, "my parents had no qualms about my getting on my bike. We had parameters; you could only go so far depending on your age. It was safe. By the time I was in junior high, I could in fact ride anywhere I wanted in town."

When I managed a golf driving range south of town one summer—I was fourteen or fifteen—I felt perfectly safe hitchhiking back and forth with money in a cigar box. When I closed late in the evening—sometimes not until 11 p.m.—I walked out to the highway, stood under the lighted sign for the Kingman Drive-In movie theater, and hitched back to town with my money box. My father and I agreed that if I didn't get a ride by a certain time, I should go back to the driving range building and wait for him. But I usually did, fairly often with the same guy, returning to Delaware from work in Columbus and driving a fine, new Nash with brightly colored dashboard lights. I felt perfectly safe.

———

That Delaware no longer exists except in memory and imagination and in local histories. Old Delaware, however, was real in a factual and verifiable sense, but I (and others) have partly invented it because we value heritage and a sense of self. Neither is a given. We select certain facts, events, and experiences and ignore others to create a heritage that connects the past with the present. Heritage establishes order and continuity in our lives. It tells us who we are. Belonging to a place—in the sense I've proposed—is fundamental to a sense of self.

Delaware enabled me, in many ways, to succeed in the larger, newer, constantly changing, uncertain world because of the ways it grounded me. But at the same time, Delaware also disabled me. It was a small, unsophisticated Ohio town of conservative values. It fit well the "Ohio" stereotype of midwestern ignorance and naïveté. I was a Delaware boy through and through with expectations that the world was simpler, more transparent, more certain, and less conflicted than it is. My high school and the town not only introduced me to racial and class differences; it also embedded in me racial prejudice and class bias. Even as an adult with, I'd like to think, a clear sense of social justice and an active record, I nevertheless live in continual tension—the residue of those old biases struggling with a clear intellectual awareness and moral commitment.

I had to leave Delaware, my youth, and my young adulthood to understand the failures of my hometown and of myself, as well as its successes. In Jim Cox's terms, I had to "wade out" in my own way in order to return many years later in memory, emotion, and imagination with a complex, ironic vision of the place that shaped me so profoundly.

14.

The Road Out

Ohio Wesleyan University

I crossed South Sandusky at Spring Street and left behind the Greyhound bus station and the post office on the two Delaware corners opposite the university and entered the Ohio Wesleyan campus. The university had been there all of my life. I was a townie. My parents were alums. I'd attended Wesleyan football and basketball games with my father. Students shopped downtown. Male students had rooms in private residences. Coeds walked through town from their dorms to the campus every day. We saw college students often. At OWU homecomings, my father entertained visiting fraternity brothers and friends at our house. So, yes, the university was always there. But when I crossed that street for the first time as an Ohio Wesleyan student, I was symbolically crossing into a new world—leaving childhood, youth, high school, and hometown behind and going to a different place.

I followed the walkway along the north side of campus for a short distance. Like other students headed to class, I turned right off the entrance walk, ascended four steps, and went on toward freshman English with Robert K. Marshall in University Hall, study in Slocum Library, college algebra in Sturges Hall, C.O. Berg's zoology in Merrick Hall, Oral English, and physical education in Edwards Gym on the south side of the campus.

University Hall, Slocum Library, and Sturges date from the nineteenth century and form the campus facade. The buildings face Sandusky Street across a wide, landscaped lawn. They expressed the separation of Ohio Wesleyan from the town and acted as a kind of barrier to the rest of the campus.

Other main campus buildings, including the following, were situated behind these three:

- Merrick Hall (a stately limestone building completed in 1873 and restored in 2015 after a twenty-five-year vacancy; I studied zoology there my freshman year);
- Elliot Hall (built as a luxury hotel in the 1830s, subsequently part of the university's original Doric front, then moved to its present interior site);
- The WSLN radio building (a "temporary" building constructed for World War II use, where I would spend countless hours);
- World War II Quonset huts, then used for classes (where I took a combined art and music appreciation course and second year Spanish, which I cut most Saturday mornings).

Only a few university buildings were located outside the campus core—Edgar Hall, at the north edge of campus, fronting directly on Sandusky; and on the Delaware side of the street, the Memorial Union Building, Air Force ROTC headquarters, and the bookstore. Women's residence halls and the music and art buildings were located blocks to the west near the original site of the nineteenth-century Ohio Wesleyan Female College.

With two campuses and students and faculty living throughout the city, Ohio Wesleyan penetrated Delaware widely. Nevertheless, the university existed, for many Delawareans, in a separate world. As students, we were embedded in the university's academic and social life whether we were on campus or not. If we were shopping at Wilson's, Woolworth's, or the Nectar, eating at Bun's or the Hamburger Inn, drinking at the Brown Jug or the Surrey Lounge (we could get 3.2 percent beer at age eighteen), or even living out in the city, we were still Ohio Wesleyan students, *not* Delawareans. Delaware had become a college town for me and no longer just my hometown.

Located just south of downtown Delaware, the main campus could have been situated miles away because architecturally, psychologically, socially, intellectually, and culturally it was. The campus gathered all these elements together in a park-like setting with clear boundaries that separated the university from the city. Ohio Wesleyan expressed, enabled, and enacted academic and social values and practices quite distinct from those of the town or even Willis High School. It introduced me to forms of life I had not known or imagined.

If I am fortunate, what happened to me in this special place will evoke your own memories of school or college—your own emergence. Our specific experiences might have occurred in different decades, at distinctly different colleges, in quite different places. What is it then between us? The power of the primary places in our lives connects us.

University Hall is an imposing, Romanesque-style building dominated by a tall bell tower—the symbol of Ohio Wesleyan since 1893. When I was there, it housed faculty offices and classrooms, as well as the president's and other administrative offices. We typically entered the building through a side entrance—climbing twelve steps to the double doors leading into a large lobby on the main floor. The wood floorboards there and in the hallways creaked, as did every stair step to the upper floors. The building smelled dusty and musty. Its age, interior, and functions expressed for me the character and purpose of a traditional liberal arts college—the timeless value of studying the arts, humanities, sciences, and social sciences.

When I walked through University Hall a few years ago, little seemed to have changed—the floors and stairs still creaked; once operating fireplaces were still lodged in a few classroom walls; the configuration of a renovated Gray Chapel remained as it had been; classes still met in the building; the president's office and other administrative offices were also there. University Hall, the place, gathered together every aspect of the university.

I remember Gray Chapel most clearly—the large assembly hall where the students attended required chapel three times a week. Student monitors checked attendance. Mondays and Wednesdays were devoted to religion and other serious matters. Friday chapel was more informal—a day, for example, when the Four Freshman performed, or we nominated candidates for campus-wide offices, or ODK, a national honor society, held its Tapping Day.

I wore a dark blue suit, pink dress shirt, and dark tie to nominate Bill Jennings for senior class president—lauding his qualifications, holding his name until the end of my speech like they did (so I thought) at real political conventions. I was wearing my khaki Air Force ROTC uniform the day ODK tapped me for membership. No one knew who would be selected. Senior ODK members roamed the hall searching for selectees, while the students tapped their feet on the wood floor. The members searched; the noise grew louder; the suspense increased. The ODK seniors surprised each selectee either by tapping him from behind (grabbing his arms or shoulders, actually) or by walking past his seat, then turning abruptly upon him. I expected to be tapped, but still I felt a rush of surprise and delight when I was.

Gray Chapel hosted concerts and evening lectures, as well. Robert Frost read; actor Henry Hull performed; editors Norman Cousins (*The Saturday Review*) and Edward Weeks (*The Atlantic*) spoke; the Minneapolis Symphony played (my first enchantment with Bartok); the Stan Kenton, Ray Anthony and Duke Ellington orchestras appeared. And that is where I first heard Dave Brubeck during his "jazz goes to college" years.

I will never forget the night Maynard Ferguson, Stan Kenton's trumpeter, stood on top of the organ, waving a white handkerchief in one hand and his trumpet in the other, bumping and grinding to the music. We were delirious.

Image 10. University Hall. Courtesy Ohio Wesleyan University.

Designed and built as the campus church, assembly hall, and concert venue, Gray Chapel was constructed as a large hemisphere with curving rows of main floor and balcony seats that focused attention on the stage, the podium, and the massive pipe organ above it. Everyone came together in that single space. It provided common experiences that enabled the university to fulfill its educational, social, and spiritual mission. The architecture, the history of the space, and its uses transformed built space into a *place*—into an expressive medium.

Slocum Library, a Renaissance-style structure, stood next to University Hall at the effective center of the campus. Virtually every student passed by it at some point during the class day and many studied at Slocum. In the second floor reading room, we sat on hard chairs at long wooden tables, lighted by fixed table lamps, surrounded by walls lined with books. A narrow balcony circled the

Image 11. Slocum reading room.

Image 12. Slocum reading room 2013.

reading room and led to offices and small classrooms. Beautiful stained-glass windows lined the balcony walls. High above, natural light flowed in through a spectacular stained-glass skylight.

As the university's intellectual center, Slocum expressed the values we all presumably shared. We were there to study and read books—even if in reality Slocum was as much social space as study space. John Reid, the librarian, shushed students over and over in futile attempts to keep the reading room for reading. The space for real study was the small first floor reserve room—except for interruptions like the time my ex-Marine fraternity brother slid a long snake though the book return slot.

I studied at Slocum afternoons and evenings—despite the distractions, such as the time my roommate booby-trapped a friend's textbook, so that when he opened it, a long paper penis stood erect. But I remember Slocum most clearly for studying and especially the time Sarah Selkirk and I sat on the library front steps, on a mild spring afternoon, and talked about John Steinbeck's *East of Eden*—a novel Professor Ben Spencer had asked his literary criticism seminar to write about. It was a difficult book for me, and I hadn't written that kind of critique before, so I needed Sarah's good mind and ideas to help me think about the novel. I remember our conversation. I have no idea what I wrote.

The university constructed a new library in 1966, but in the early 2000s it restored the Slocum Reading Room for student study to look exactly as it did in my day. I'd not known that when I visited a few years ago, so I was momentarily confused. My memory, archival pictures I had just seen, and then the actual room before me all ran together. I caught myself looking down from the balcony trying to see myself and my friends.

In the 1950s, Ohio Wesleyan was a separate and protected place where young men and women stepped aside from life into a liminal space between adolescence and adulthood while they studied and learned and prepared for their futures. Many of my high school classmates, by contrast, were married soon after high school graduation, assumed full-time work, and started families. Fully embedded in the life of the city, they were true Delawareans—real people beginning their adult lives at age eighteen or nineteen.

The landscaped space, academic buildings, and residence halls of the Ohio Wesleyan campus provided the stage for this interregnum. It was a separate place that expressed the university's character and purpose in that separation, as well as in the architecture and setting. As a place, the buildings, the campus, the faculty, students, and the activity of the university were inseparably intertwined. The place was both a presence and a force. It not only told an eloquent story of the past and present of the university, but it also shaped its students' identities and influenced the rest of their lives.

I entered Ohio Wesleyan with no specific goals for my future and no sense of what a college education could be. I joined a fraternity. I played junior varsity basketball and varsity baseball. I took required courses. I was waiting for something to happen.

During my sophomore year, I lived at the fraternity house, where I shared a third floor room with two others—Jack McKinnie, a friend from high school, and Roger Fromm. We kept our books, clothes, and personal belongings in the room, but we slept in one of the two, second floor dormitories—large spaces filled with double-deck bunks where all house residents slept. Ours was a plain room with three desks and chairs, closets, a worn rug (I think), and an old sofa tucked in the large dormer facing the rear of the house.

When I reflect for a moment on this room and on the sleeping areas, the dining room where the brothers shared meals at designated times, the living rooms at the front of the house, and the secret chapter room in the basement, I realize that such commonplace spaces carry thick social and cultural meanings—even if their ordinariness masks their expressiveness. The spaces expressed and enabled a fraternity form of life. Fairbanks Lodge told a story of the past—my father lived there when he was at Ohio Wesleyan—and of the present I was living.

The brothers' singing concentrates for me the entire fraternity experience. At meals, singing might begin spontaneously. At other times, our leader, Stu Root, conducted us. Communal singing not only brings people together in harmony and joy, but singing in unison can also synchronize the breathing and even the heartbeats of the group. In voice, spirit, and physiology, the brothers were one. I've been moved recently by the deep sense of brotherhood my aging fraternity brothers express. "Fraternity" is fundamental to their lives—as it has never been to mine.

My social life did center on the fraternity for a few years. There were Sunday date nights, special parties, and also all-campus dances. My pledge brother, Bill McAfee, introduced me to Mickey Stahl when I was still a freshman and she a sophomore cheerleader.

Later in the spring of that year, we began seeing one another regularly, and still later became pinned, engaged, and eventually married. During the summer my father was dying, she lived with us for weeks to help care for him. She is the mother of the four daughters to whom I've dedicated this book.

Basketball and baseball provided another kind of social life—the society of teammates—as well as sport itself. I lasted one year playing basketball; I pitched varsity baseball for two. Neither sport, however, gave me the sense of team that high school athletics had. The basketball and baseball teams at Willis live in my memory as true teams. But I had to search the campus newspaper for the names of my college teammates, and only then did I remember who they were.

At mid-season, I became the starting center on the JV basketball team. I played game after game—as I often had in high school—against opponents four to six inches taller than I. I tired of that. I tired of an ignorant coach saying over and over, "Fred, you've got to score more"—without telling me how. I did manage twelve points in one of the late games of the season. I might have made the varsity my sophomore year, but I didn't try. I was done with the game.

During my sophomore baseball season, the team lost twice as many games as it won. It was a discouraging time. I turned out to be a mediocre, small-college pitcher. "We could do all right if our pitching improved," the coach told the campus newspaper. The coaches themselves were also mediocre and uninspiring. One (also the freshman basketball coach) told me repeatedly, "Fred, you've gotta make the ball break more"—without telling me how. Something was wrong. I wasn't enjoying the game as I had in high school. Moreover, my father—my greatest cheerleader—died the following September. I lost that incentive for continuing in sport. I realized, as well, that I didn't depend on athletics for a sense of self. Nor did I find in fraternity the brotherhood and meaning others

did. I had come to an important turning point. It was clearly time
to move on.

I enrolled at Wesleyan as an engineering major. My paternal
grandfather was a self-taught engineer and designer of glass-
making machines. Engineering seemed to make sense. I had no
other inclinations—even though I had always been known as a
reader, as a boy of words, so to speak. I had done well, however, in
Raymond Felts's high school math classes, so why not engineering? One semester of college algebra convinced me otherwise.

I took Freshman English from Robert K. Marshall—or "Bobby
Marsh," as students privately called him. He was a warm, bright,
attentive man who spoke with a soft North Carolina accent. A few
years earlier, he had published *Little Squire Jim*, his first novel,
which the *Saturday Review* described as "a story of extraordinary
vitality and of depth sufficient to invite the attention of the most
discerning fiction readers." It is about the people "who inhabit the
remote reaches of North Carolina's mountains." My teacher had
become a celebrity in Delaware, especially because of the mildly
erotic parts of the book.

Until Professor Marshall told me—in his kind and encouraging
way—that my writing was awkward, wordy, and unfocused, I had
no idea. My high school senior English teacher never told me that.
Even so, I managed a B, and by the end of the year, apparently, I
wrote enough better to earn my first A in English. It would be a
long slog, however, through Advanced Comp with Ruth Davies
(a B- semester), paper after undergraduate paper, poorly written
graduate essays (at Indiana University, a professor once described
my style as "bizarre"), and years of teaching writing before I wrote
with any skill or effect.

I liked Professor Marshall, and apparently, he thought well
of me. So I enrolled the next year in his survey of British literature. Uncertain about a major and vague about my future, I'd
been taking mainly liberal education requirements—history,

political science, Spanish, Bible, philosophy, physical education—the common intellectual coin so valued at Ohio Wesleyan.
Professor Marshall's course decided it. I declared an English major
and entered a world that would transform my life and give me a
career—although at the time I certainly didn't know it. I knew
only that I liked reading books.

I had also enrolled in the introductory courses for a radio and
television major. After I'd spoken at my high school commencement, people complimented me on my voice and speaking style—
"You ought to go on the radio," a few said. So I decided to try.
I appeared in my first radio play, "The Haunted Typewriter," on
WSLN in October of my first semester. That spring, I became a
Tuesday and Thursday morning disc jockey. I selected the music,
introduced each tune, and then signaled the engineer on the other
side of the glass to play the record.

I made my first radio mistake when I comically (or so I thought)
garbled the name of a popular French singing group. On the
second or third note of the song, Stuart Postle, our faculty adviser,
stepped into the studio and graciously but firmly said, "Fred, if
you don't know how to pronounce a name, learn how. Don't ever
do that, again." I didn't.

I also learned the next year that music—and not my own com-
mentary—was the point of my nighttime show. I had thought
of myself as a radio personality like the DJs on Columbus and
Cincinnati commercial radio. "Last night, I played only ten tunes
on my one-hour show," I once boasted to a friend. Obviously, I
had talked on and on about music and artists the rest of the time.
One day a fraternity brother took me aside and said kindly, "I like
your show, Freddie, but I listen to it for the music. You should
play more." And I did—no longer a radio "personality." These were
small but important corrections—perhaps humiliations—that I'm
guessing we've all experienced in some form.

WSLN began broadcasting in February 1952—six months
before I enrolled at Ohio Wesleyan. It was housed in a small

wooden building constructed during World War II for V-5 and V-12 cadet use. Located in the center of campus, there was one classroom, two offices, a small studio for the staff announcer, a larger studio for programs, and the engineer's console and turntable room with windows into each studio. It had bare wooden floors, flimsy walls, and studios soundproofed with egg cartons. The station broadcast with a weak FM signal. The building suggested that WSLN was a minor undertaking for the university, but it was vital to my life. As a sophomore, I served as popular music director, then as program director the next year, and finally as station manager my senior year.

Stuart Postle dressed handsomely in tailored suits and silk ties—and sometimes for class, he wore a light sport coat with a black shirt buttoned at the neck. In winter he wore a long, dark top coat and white silk scarf as he walked to campus from his home on North Franklin Street, leading his light brown spaniel on a leash—a rather vicious dog, actually, that no one at the station dared pet. Stuart smoked with a long cigarette holder, spoke with a rich, cultivated radio voice, traveled to New York during holidays to see shows and then gave "splendid" accounts to Delaware service clubs of the plays he'd seen. "Splendid" was one of his favorite words. He was a handsome, stylish single man—elegant and rather theatrical but in ways we all loved. Students fantasized about a lost love—a woman who had left him or died, someone he could not replace. In those days, we didn't really know. He left Ohio Wesleyan before my senior year—delighted, I'm guessing, to escape Delaware.

I learned from him—not only about singing groups but also about proper radio speech and articulation. It's record, not rekurd; ed-you-cation, not edjication; double U, not dubya; just, not jist; and so on. I was so taken by ar-tick-u-lat-ting every syllable properly that in Ben Spencer's English class one day, I added a hard *c* to Connecticut—Connekticut—much to the loud amusement of the professor and the entire class. I also learned from Stuart how to manage and lead, but I value most the conversation we had when my father was dying.

Eight years earlier, my father left public school administration and opened the "Personal Insurance Service." Even though I was majoring in English and in radio and TV, I had been wondering whether I should change my major to economics or business, so that I could take over my father's business. I was being unrealistic. I had two years of college left—perhaps more with a change of major—and my father would die within months. Still, I felt some obligation. I explained all this one afternoon to Stuart. And then very skillfully and considerately, through questions and conversation, Stuart helped me understand that I was not obligated to do this, that I did not really want to change majors, and that I did not want to live in Delaware after college. That conversation gave me my first sense of what I really cared about.

Three photographs stand on the desk I inherited from my mother—my father, Ervin; my mother, Winifred; and Benjamin Townley Spencer, my professor, mentor, and intellectual father—and then later a dear friend.

Ben Spencer taught American literature, Shakespeare, and Literary Criticism, and he published in each area as well. Critics described *The Quest for Nationality*, one of his two admired books about American literary nationalism, as "One of the books you have to read and digest at the peril of not knowing the field. It has in its field no peer."

I took all but one of his literature courses. I did not do well at first, for I needed time to understand the major and develop intellectually. My junior transcript shows more Bs than As. But Ben encouraged me when I expressed some frustration and doubt about my ability. "You'll get your As," he said reassuringly. And I did, from him, as well as from other professors—especially during my last three semesters—my final one being my only 4.0 term.

Ben lectured from a small notebook filled with his tiny script— the same miniscule hand he wrote with on papers and exams. A story circulated that when a student asked him to decipher a comment he had written on a paper, Ben read, "You should try to write more legibly." I tried to emulate his handwriting, and over the years, I succeeded so completely that friends and students sometimes couldn't decipher my script—nor occasionally could I. Ben lectured smoothly, skillfully, and knowledgeably, with sensitivity, enjoyment, and wit, and with a slightly bemused smile. His was a quiet but powerful presence. Often and without warning, he'd call on a student by name to agree, comment, or wake up. "Mr. Carlisle, why does Whitman assert at the end of *Song of Myself,* 'Do I contradict myself / Very well then I contradict myself'? What is there about Isabel Archer that seems so American?"

It was a delight to sit in his class and listen—and sometimes comment. "Students looked forward to going to his class," a friend observed, "*if* they were prepared." Ben was demanding but warm and engaging.

The photograph captures him beautifully. He's wearing a tweed suit with wide lapels and a knit tie; he is seated on a sofa, his right arm resting on its back, a pipe in his hand; his other hand rests in his lap. A balding man wearing glasses, he is smiling. Two folders and a book lie next to him on the sofa; a stack of books sits on a table behind. The picture reveals his quiet but powerful presence and suggests his gentlemanly and courtly manner.

Ben seems at home in the photograph, as he was truly at home at 110 University Avenue. The traditional Georgian architecture and the house as living and hosting space expressed him well. In Bachelard's terms, the house "revealed the topography of [his] intimate being"—of his domestic and intellectual life. He made the space come alive residing, reading, writing, and teaching there. He and Virginia transformed a space they had acquired into a place rich with experience and meaning. It was traditional in design, enduring in its uses, and thereby appropriate for a classic, even the ideal, university professor. It was a beautiful example of

Image 13. Benjamin T. Spencer. Courtesy of the Ohio Wesleyan University Historical Collection.

place as a complex event. Ohio Wesleyan razed the house some years ago, but like Ben and his classes, it remains not only a memory but also an effect.

In American literature, we read the canon of the time—plodding through writers like Bryant, Whittier, Longfellow, and Lowell, and reading Hawthorne, Emerson, Thoreau, and even Melville—with interest but no real passion. It would be some years before I understood *Moby Dick*—its comedy, as well as its drama and philosophical explorations—and read and taught it over and over. The senior who assisted Ben by grading my book report about *Moby Dick* remarked with some puzzlement, "I'd hoped you'd see more to it than that."

But then I discovered Walt Whitman. His long, passionate, celebratory lines captured my imagination.

> I celebrate myself, and sing myself,
> And what I assume you shall assume
> For every atom belonging to me as good belongs to you.

Finally: a writer to care about emotionally. Ben smiled knowingly when my friend Lloyd Gardner and I explained our response to *Leaves of Grass*. Weeks later, I discovered Henry James, a vastly different writer but one who also appealed to my imagination. I was fascinated by his portrayal of the cultural encounters between the new and old worlds. Years later, I planned a book about the two—both quintessentially American writers. Nothing came of it, however. But I did write about Whitman and *Leaves of Grass*.

Lloyd Gardner and I grew up in Delaware at the same time. We didn't know one another well in elementary school. He attended West and I North. We became good friends in high school, as well as occasional rivals for roles in school plays. One year, unable to decide, the drama teacher cast us both as the father in *Cheaper by the Dozen*. Lloyd played the role one night and I the next. We shared an interest in drama. We were both good students, but our school worlds also diverged. Lloyd was an accomplished debater at Willis and a superb one at Ohio Wesleyan. I played basketball and baseball and was active in student government.

In the spring of our junior year, Ben invited Lloyd and me to assist him the following year by taking book reports from his American literature students. Each semester, he distributed a list of books and required students to read a certain number of them in addition to their daily assignments. Students then reported orally to Lloyd, me, or Ben. In mild weather, we often sat on the fire escape behind Slocum Library. In winter, we used Ben's office or a vacant one nearby. We asked questions; students answered. We informed Ben of the results. Lloyd and I took our responsibility quite seriously. We also experienced a real ego boost. We'd been recognized and approved by Ben Spencer—department chair, superb teacher, and published scholar. This was my first practical move toward a university life.

Ben Spencer, in the end, taught me how to be a professor and, as my father had as well, how to be a human being—even if I have fallen short of their models. Throughout my life—as the picture on my desk suggests—Ben Spencer has remained a powerful presence and influence.

Years later, I tried to acknowledge that influence and express my gratitude by dedicating my Whitman book to him and also by arranging for him to teach one summer in the Michigan State English Department where I was a faculty member. In 1977, in a publication announcing the Benjamin Townley Spencer Professorship in Literature, I said at the end of my remarks, "Benjamin T. Spencer comes as close to the ideal teacher and scholar as most of us will ever see."

Ohio Wesleyan broadened me intellectually and culturally, to be sure, but it also narrowed my experience and limited my life to a career inside universities, isolated from the diverse society I'd known in high school and that characterized Delaware. Ohio Wesleyan was mainly a white, middle-class college. The six blacks in my graduating class constituted less than 2 percent of the whole. I had very little contact with them—except for Russ Davis on the JV basketball team—but I had no contact with him after that season. Otherwise, both baseball and basketball players were all white. No Blacks were involved in the radio station, as I recall, and I don't remember any as English majors. I knew who the Black students were. I could smile and say "hello." But I didn't know them in any sense.

The fraternity certainly was all white. Even though the university was in the early stages of anti-discrimination policies, the Greek organizations—and most white students were members— engaged in "subtle, yet widespread discriminatory practices," according to a university history. In a dramatic public gesture in 1952, Frank Stanton, the president of CBS and a Wesleyan graduate, resigned from his fraternity because of its discrimination clause for membership, requiring "pure Aryan blood."

As educated and changed as I was, I left Ohio Wesleyan still a rather naïve and ignorant, small-town Delaware boy. I graduated in June and went directly to Ohio State University for master's study. There, accomplished graduate students marveled at my mispronunciations of Albert Camus's name and of Townshend Hall. I enunciated the letter s in Camus (probably the t in Albert as well) and the sh in Townshend. In a seminar on Metaphysical poets, where each student read aloud and then explicated a John Donne or other metaphysical poem in great detail, I felt (and was) inadequate. These poets were new to me, and the approach to exegesis was as well. On the whole, I performed adequately at Ohio State. Two faculty members encouraged me to stay on for a PhD. They thought I could probably do it. But I wasn't ready. James Logan, my thesis director, said my thesis was better than he thought it would be. Hardly a compliment. Professor Andy Wright, a great supporter, hoped I could learn to write as well as I could speak. It was time, however, to serve other obligations—mainly the United States Air Force—and come back to academia later.

———

Ohio Wesleyan—this distinct and separate place—expressed values I came to understand. It enabled me to begin discovering my adult self, a new expression of the Delaware boy who went to college; and it enacted the change. Ohio Wesleyan transformed my life and not only gave me a profession; it also showed me forms of life I had not dreamed of. Besides its social and cultural influences, the university also opened into a vast, unending abstract space. It introduced me to another kind of dwelling—an intellectual and imaginative place that has centered my life for decades.

Ohio Wesleyan opened a road for me toward reading, reflecting, imagining, teaching, and writing that would in time enable me to rethink what I'd rejected and recover what I'd lost. In some sense, I dwell in these capacities and with the people, places, and heritage they've opened for me. They give me eyes to see, ideas to guide me, and ways to imagine and express my discoveries. What seemed to separate me from my hometown and from my past has actually given them back to me.

Place is not only a location and built space; it is also psychological, social, cultural, remembered, and imagined space. People, actions, and spaces intertwine into a complex, inseparable whole and into a place of power. Ohio Wesleyan conditioned the way I have lived and shaped who I have become. Even though it was located a few blocks south of downtown Delaware—near the city's center—and about the same distance from downtown as Willis High School, it was not a center of Delaware City the way Willis was. No physical walls or fences surrounded the campus, but its borders were clear and its separateness real. The walls were psychological, social, and cultural.

Psychologically, the university provided a protected place for young men and women to pass from late adolescence into their adult and professional lives

Socially, I entered a more cosmopolitan society with students from large cities, other states, even other countries, and from families more affluent and traveled than mine. I was also meeting university professors—a class quite different from public school teachers. I experienced difference but in a relatively comfortable and somewhat familiar way. In contrast, my life narrowed socially even more, further isolating me from blacks and people from other social and economic classes. My life would stay that way for decades.

Ohio Wesleyan's mission and culture fit more closely with liberal arts colleges across the country than it did with Delaware's small-town life. The university expressed, enabled, and enacted an upper middle-class, predominantly white, Protestant form of life at the same time it advanced an intellectual life that sometimes subverted the very class it served.

15.

A Moveable Place

I could have returned to Delaware after graduate school. Ben Spencer asked—indirectly, to be sure—if I might come back to Ohio Wesleyan to teach. I could have misread him. It's so easy to misinterpret what is happening to us—or what has happened. But I declined Ben's overture—also indirectly. I was feeling my greatest alienation from Delaware at that point. I had to escape its small-town character. I had to find a future elsewhere—out there, far beyond the Central Ohio horizon. Delaware was not in my future.

I left my hometown in almost every sense and went on to a relatively placeless life. I was a man always leaving his pasts and moving on to the next opportunity—typical of the college-educated, American, middle-class culture. Edward Casey describes it as "an insidious nomadism endemic to modern times, in which the individual . . . drifts within indifferent spaces of housing developments and shopping centers and superhighways." Many of us experience this—even if we are unaware—as I certainly was.

I studied at two universities for graduate work; I served in the U.S. Air Force at two different bases; I have had appointments at five universities; I worked with the federal universities in the United Arab Emirates for ten years; I've lived in ten towns and cities; I have visited many countries; I have been married three times. I developed, learned, and matured, but no location, except for that special place in the Virginia mountains, affected me in the way my hometown and my college did. I did not truly *dwell* in any of them. I simply kept moving from one university and college town to another, living in similar but not identical situations. I spent most of my time with university faculty and students—different individuals but common types.

I was living in a moveable place. Like a true nomad, I was carrying my place with me from location to location, situation to situation. I was living within an academic culture—a deep, guiding set of interests, behaviors, and values that remained relatively constant from one university to another. I'd transformed the cultural forces of academic life into meanings that I carried with me from one location to another. This cultural place—the indwelling force—was based on abstraction and was therefore moveable. It was also based on concretization. That is, in each place the culture manifested itself in specific interests, objects, and actions, expressed both professionally and personally.

I have carried my art collection and library with me from one residence to another for decades. They reflect not just my interests and tastes. They express something of the person I am and capture a part of my becoming over time. They represent the culture I have carried from location to location. I dwell with them, and they dwell with me. I bought my first piece of art—"Urbino Notte" by Donald Saff—in 1964. It speaks for the whole of what I own. "Urbino Notte" hangs above my Virginia writing table. It is a 28"x 24" etching and aquatint. (The Museum of Modern Art in New York owns a similar Saff print titled "Rising Sphere.") A deep-red full moon is rising over horizontal black and charcoal branches and sticks. The moon rises above the earth. I am drawn to it every time I walk into the room. I carry a mental image with me everywhere I go.

"Urbino Notte" has been with me and I with it for fifty years. I have carried it from house to house, job to job, through marriages, the raising of children, a divorce, the death of a wife, and remarriage—in short, through most of my life. It possesses a meaning and presence beyond its materiality and artistry. It even reminds me of my own days in Urbino, Italy, decades ago.

Barbara and I had arrived late Sunday afternoon with traveler's checks and only a few lira—not enough to buy food or eat in a restaurant. No banks were open. It was years before ATMs and debit cards. We explained to a policeman in the central piazza and asked what we might do. He didn't know. An Italian man, who had been talking with him, stood back and listened to our story. After a few minutes, he stepped forward and said, "I will cash your checks for you." He was from Rome and spoke English well. He'd been, we learned, a POW in the United States during World War II. We were puzzled—an individual offering to cash traveler's checks? "But how can you do this?" Barbara asked. "Ah, Signora, in Rome, anything is possible."

Walt Whitman's *Leaves of Grass* speaks for my books the way "Urbino Notte" does for art. I discovered Whitman in Ben Spencer's American literature class; I rediscovered him at Indiana University with a former student of Ben's and my graduate mentor, Edwin Cady. I taught the poetry many times, and I have written about Whitman. Years ago when my reading and working copy of *Leaves of Grass* began to fall apart, I placed it in a three-ring binder. Frequent handling and notes scrawled throughout, as well as the age of the paper, had caused the pages to wear and even tear. The cover had disappeared long ago. I keep that copy in a plastic bag to preserve what little remains of the object—paper, print, words, lines, and poems. But when I hold it and listen or read the poems in another edition, Whitman is there—in all his passion, energy, richness, and beauty.

He speaks boldly—"I sound my barbaric yawp over the rooftops of the world." That's the marvel of Whitman and of all great poets. The voice and presence emerge from paper and print. In Whitman's case, he leaps from the page and grabs us or quietly seduces us into his presence.

> Whoever you are holding me now in hand,
> Without one thing all will be useless,

I give you fair warning before you attempt me further,
I am not what you supposed, but far different.

And in his bravado, he would speak for us all—and know us better than we know ourselves.

I celebrate myself, and sing myself,
And what I assume you shall assume,
For every atom belonging to me as good belongs to you.

Whitman is big and he is bold. He pushes himself in front of every other book I own—and thereby speaks for them all. He is a massive presence. I imagine all the other authors—all my other books—standing beside or behind him waiting their turns to speak.

My books, prints, paintings, ceramics, and sculpture surround me. They inhabit every room. They provide not only continuity but also order, stability, comfort, domesticity, even a kind of intimacy. They have made me feel at home wherever I live.

I have taken with me a personal culture as well—both as an abstraction, grounded in behavior and values, and manifest in relationships—a culture of marriage and children. When a future sister-in-law asked a few years ago, "Why do you want to marry? Why not just live together?" I tried to explain. "Marriage has been a way of life for me virtually all of my adulthood. And changing cultures isn't easy—especially when there's no good reason. Marriage makes my love and commitment public and clear. It shows greater respect, I believe, than simply living together." I doubt that I convinced her, but that's the personal culture I've lived within for decades and have moved from location to location.

16.

_____

New Delaware

The Place Is Still the Thing

I do not live in Delaware, Ohio. Nevertheless, I have "refound" my hometown. I've been *placed* there in a way I'd never been before. I had to leave—and Ohio Wesleyan opened the door—in order to return and know the town where I grew up. I had to rediscover James M. Cox and at the same time pass through Clover Hollow— after decades of a nomadic life—to understand the reality, value, and power of place and past. The Hollow and the man allowed me to discover *for the first time* my primal place—208 West Lincoln Avenue—and the centers of my young life that shaped me so profoundly: my Lincoln Avenue neighborhood, the woods, North Elementary, Willis High School, and downtown Delaware. I am placed now historically, emotionally, psychologically, socially, culturally, and imaginatively—if not physically—in my hometown.

The Delaware I've recovered is not a static place fixed in my memory or set by history. The hometown of my birth, childhood, youth, and young adulthood was in itself dynamic, and it changes with every memory and reinterpretation. Every time my friend Lloyd Gardner and I talk about our Delaware, it changes in some sense as we remember or read about moments we had not known before. I emceed my fiftieth high school reunion. We reminisced. We also talked about my classmates' recent histories and present lives. For most of them, Delaware extends over time. Theirs was a sixty- to seventy-year Delaware—a dynamic, changing city.

I visited Delaware while my mother lived there, where Ben Spencer lived until his death, where an elementary school carries my father's name—the Ervin F. Carlisle Elementary School. I have also been returning regularly to the New Delaware—as I think of it—for the last ten years to search for my father, to renew a few friendships, to establish an endowment at the university, and to understand how Delaware has and has not changed. The New Delaware is also part of the hometown I have recovered. The city has been transformed, yet in a certain sense it is the same place—my hometown and not just in memory and history but also in the present at this very moment.

Within a few decades, the centered, safe, largely self-sufficient, yet flawed city where I grew up began to change dramatically. The downtown declined. The economy stagnated. The city suffered losses similar to those in small towns and cities across the Midwest. "Downtown pretty much ceased to exist in the late '80s and early '90s," Roger Koch, a local developer observes. The box store revolution began. Wal-Mart opened, Kroger built a large store south of the city center, and a strip mall opened in the same area. "Within ten years nearly every locally owned business that had been downtown ceased to exist. It was pretty grim. The downtown looked awful," Roger said. "We had a 50 percent vacancy rate."

No businesses I knew survived, except for three financial institutions, several legal firms and insurance agencies, the Strand Theatre, a few bars, Bun's restaurant, and The Hamburger Inn. The groceries and meat market; clothing and shoe stores; drug, hardware, and department stores; the ten-cent stores; retailers like Sears; and automobile garages and automotive sales all moved or closed.

The city's industrial base also declined sharply. We lost "hundreds and hundreds of jobs in the '60s, '70s, and '80s," according to Jack Hilborn, a friend and life-long resident and businessman. The Hughes-Keenan Corporation, for example—a major employer and large manufacturer of van bodies, Ford chassis, and hydraulic lifts—went bankrupt in 1985 and ceased operations in January 1986. After being bought out, the Sunray Stove Company left the city and eventually ceased operations. The Ranco Corporation, an electrical equipment and parts company, closed its Delaware plant. Grumman Flexible, a manufacturer of city transit buses, built a large factory in the 1970s but then closed it in the early 1980s. Hundreds of jobs lost.

These losses undermined community spirit and threatened the city with a permanent downward spiral. Delaware, however, arrested the decline, and unlike many small towns began a long renewal. When I visit Delaware now, I return to a city of roughly 38,000 people (four times as large as when I lived there), to a renewed downtown, to a city redeveloping its business and industrial base, to a place trying to sustain itself as a community with values similar to the Old Delaware. I could see this, but I didn't know what the city had done to reverse the decline, invigorate the town, and assure it of a sustainable future.

To help answer my questions, Jack Hilborn gathered a group of civic and business leaders together—including the city manager, a county commissioner, a property developer, a bank executive, a prominent business leader, and the president of the historical society. There was one woman in the all-white group.

We met at the Delaware County Courthouse—a handsome, 1868, redbrick, Italianate structure that dominates the block just north of the downtown. The building captures Delaware's and the county's one-hundred-fifty-year history—literally, in the records on file and, symbolically, as a center of government and authority. I was meeting with established leaders in a place of tradition and authority. A few months later, I would meet in very different kinds of places with black leaders in Delaware's South End.

The courthouse group focused on leadership, location, and institutions. This New Delaware did not just happen, they explained. It took determination and collaboration—"twenty years of concerted effort." I learned about the city's decline and recovery from the meeting, but more than that, it made me proud of my hometown. Nevertheless, I knew that while this was the dominant story, it was not the whole story.

"I moved to Delaware thirty-two years ago," Roger Koch, an architect and local developer, explained, "because it looked like a very promising place to live. Even in the 1970s, I could see that Columbus and Central Ohio had one of the few positive economies in the Midwest and that the economic effects of Columbus were going to move north and northeast"—into Delaware County and toward the city proper.

"So I began to look at downtown Delaware as an opportunity in spite of the 50 percent vacancy rate," Koch said. "My brother and I bought one building for a bargain price. We bought that one and rehabbed it. We rented it to a different user. Then we did another one and another one—and mostly with borrowed money."

The city assumed an important role, first by obtaining a state grant for downtown redevelopment, then by borrowing money to supplement the grant, and also by levying an assessment on downtown businesses. With these funds, the city built new sidewalks, installed new sewers, and improved the streets. According to Koch, these 2002–3 projects "showed that the city supported development. Fixing up buildings and upgrading the infrastructure made Delaware look better and improved your chances for long term success." The city, private developers, and financial institutions were collaborating to renew downtown Delaware.

The downtown, once again, is becoming the place to go. "We knew we had to turn it into a dining and entertainment center because the department stores [and all the other Old Delaware businesses] were not going to come back," said Koch. Main Street Delaware, a nonprofit volunteer organization that promotes economic development, has promoted the renewal and sponsors downtown events like First Fridays—a monthly event that focuses on specific themes such as March for Art, Picnic with the Cops, Artful Spaces Tour, Fire Prevention, free children's activities, and late store openings. "But in fact," Jack Hilborn points out, "Delaware is alive every Friday night. We have a movie theater that will soon be a hundred years old and that still shows first-run movies. We have shops, new restaurants, and galleries. Delaware has reinvented itself."

Ana Babiasz, the CEO of Fidelity Federal and a relative newcomer, has found a home—a real sense of place—in Delaware. "The people have such a commitment. It is a lovely place. It is a real place." She grew up in Philadelphia in a house and neighborhood now gone. "I cannot go back to my hometown. But you can go back," she says. I wasn't sure if she meant that in general terms or whether she was speaking to me when she said it.

"Delaware still retains its small-town character and charm," Tom Homan, the city manager suggests. "When you walk downtown, it feels like a real downtown." The population growth and the push of Columbus northward into Delaware County have turned the city into a bedroom community for some, but as the city manager believes, "Delaware is *not* a suburb. It is a full-service city." A community survey showed that people have a high regard for Delaware. They want to see a healthy downtown with more shops and restaurants.

A huge residential, commercial, and corporate development in Polaris Town Center and Polaris Fashion Place in south Delaware County epitomizes the northward move of Columbus. This is the kind of growth Roger Koch envisioned in the 1970s when he moved to Delaware. The city itself is moving south. Its corporate limits now reach well beyond the town I knew—even past Kingman Hill where I managed the golf driving range and hitchhiked home late at night. Residential development now extends to the new city limits and on toward Columbus, as well as to the west and east of Old Delaware. Grady Memorial Hospital, now part of OhioHealth, has constructed a large outpatient facility five miles south of the city center. Columbus State

Community College has built a Delaware campus just south of the city boundary.

The Olentangy School District, once separate, small, and far south of Delaware, is now part of the city. Since 1990 the district has grown to a total of twenty-three schools—fifteen elementary schools, five middle schools, and three high schools. "The Olentangy School District is now just booming," the Delaware city manager said. "Knowing we were annexing another district, we worked on ways to get them to associate with Delaware. Teenagers are now coming here."

There is virtually no break in commercial and residential development along U.S. Route 23 (Columbus Pike) between Delaware and Columbus. Once upon a time, miles of open farmland separated the two cities. Now when I drive that corridor, I see few open fields and virtually no farmland. Even the few undeveloped sites typically have signs promising future development. Delaware and Columbus are growing closer together geographically and economically, yet it takes longer to travel from one to the other than it did even twenty years ago.

Delaware has also been rebuilding its business, industrial, and service base. As one of the few industries not to leave, PPG Industries recently celebrated a fifty-year presence in Delaware. Several other companies stayed as well, so the base did not collapse altogether. A Kroger distribution center east of town with more than one thousand employees is relatively new, as is the JEGS Corporate Distribution Center. It opened in Delaware in 1999, and its owners speak glowingly of the city:

> The Delaware community continues to grow and expand in all areas; housing, schools, and business. This has helped JEGS attract a great workforce to service its customers and [ship] over one million packages annually. Delaware has become quite a home for JEGS.

A planned extension of the Saw Mill Parkway from Columbus to Delaware's industrial park will stimulate even further industrial development.

Delaware is the county seat. The courthouse and county offices are located on North Sandusky Street, just north of Central Avenue where downtown Delaware begins. The city's school system has persisted and grown. The school named for my father, for example, is undergoing an extensive remodeling and expansion at the time of this writing. Delaware is the home of Ohio Wesleyan University—a highly regarded liberal arts college. The loss of a daily newspaper in many towns has signified decline, but *The Delaware Gazette* continues to publish a daily newspaper, as it has since 1818. The existence of these strong and continuing entities and the restoration if its business and industrial base helps explain how the city has avoided the fate of other small towns and cities.

Delaware has also profited from its location near the state capital. "We have the best of both worlds—a great community and also the big city forty-five minutes down the road," Jack Hilborn says. "My generation—fiftyish—is settling here because the job opportunities are such that you can have a job in Columbus and live here in Delaware," Tom Homan added. "The city is an attractive place to live."

During the low point in Delaware's fortunes, young families were not moving there. The nice homes—a three-story, three-thousand-square-foot Italianate house, for example, would sell for $75,000 to $80,000. Now some of these same homes sell for almost half a million. "When young families are looking for a safe, small town to live in, we look really good," Roger Koch observed.

Located in a region of dynamic economic and demographic growth, stabilized and sustained by the university, city and county governments, healthcare institutions, and the city's schools, and led by committed, energetic, and inventive citizens and officials, Delaware is a classic example of small-city renewal. Collaboration, imagination, and initiative have made the difference. The change has not been by chance, they all agreed.

There is still more to do. Downtown is a continuing project. I noticed a number of vacant storefronts recently. The site of Wilson's clothing store, for example, where I worked briefly, is empty and looks rundown. There are antique stores, beauty salons, a barbershop, and a tanning salon—businesses typical of a downtown in transition. There are also, however, a number of retail specialty shops, new restaurants, galleries, and many professional offices. I could see how the city was re-centering itself.

The East Side and the South End, however, have benefitted less from Delaware's re-invention. Race and class divisions persist in the New Delaware, just as they did in my town. I drove through the old East Side on a bleak November day and then later on a sunny day in May. It looked, both times, much as it did fifty years ago. It is an area of older, modest houses, some of which have not been well kept and some of which are over one hundred years old. There are few new buildings. My initial impressions were confirmed when I talked with Mary Jane Santos, a resident and former city council member.

The population consists mainly of blue-collar, conservative whites who have lived in their homes for decades. "I can count the black families on one hand," the former councilwoman said. East Siders shop and work in Delaware. They do not typically commute. They are embedded and independent. They do not, however, identify with the city as a whole—as their typical opposition to new tax proposals suggests. They prefer, in fact, that Delaware leave them alone—unless something affects them directly. In those cases, they speak out loudly.

This is the oldest part of Delaware. It is on the other side of the river. The Olentangy, flowing south through the city, separates the East Side from the rest of the city. ("It's always been the East-West divider," someone observed at the courthouse meeting.) Then, on the eastern edge of the neighborhood, a railroad embankment—"the trestle"—divides the area from new residential and commercial developments just beyond.

The residents are isolated geographically, as well as historically and by class. They are loyal to the East Side, and they value their independence. More people own their homes there than elsewhere in the city—a fact which somewhat surprised me and doubtless reflects my own biases. "We see ourselves as permanently residing in our homes," the former councilwoman said. "My neighbors have been my neighbors for twenty-five years. The one across the street is living in his parents' home. My son bought a house eight houses away. He chose to live on the East Side."

East Siders keep to themselves by choice, at least in part. History, geography, and economics also contribute to their relative isolation. Choice and circumstances explain the static, under-developed character of the area. Even if residents or developers wanted to build or remodel, there is no room for new construction, no place to grow within that bounded area. To build, say, a new house would require tearing down an old structure. People either cannot afford that or their sense of place gives them no reason to.

Residential and commercial development has jumped over the old part and formed a new East Side just beyond the trestle—leaving the old East Side behind and, depending on your point of view, continuing its isolation and marginalization or protecting its independence. Several hundred houses have been built in the new residential developments. Most buyers moved there from the outside, many from the Columbus area. They do not feel the same attachment to the old East Side "that others of us do," Ms. Santos said. They are younger and are changing the demographic of Conger Elementary School—East School as I knew it. They are also changing the voting patterns. Their effect on long-term change remains to be seen.

The new commercial area beyond the trestle includes a Meijer superstore, Kohl's, OfficeMax, and other typical franchises. A new outlet mall is opening a bit farther east, just off Interstate 71. The East Side might benefit from the shopping, but I doubt many traditional residents will move into the new subdivisions. The old East Side remains.

My wife, Beth, and I drove into the South End on a cold, wet May morning. It's an area where most black Delawareans still live. We wanted to learn more about this part of New Delaware, "Delaware—the good, the bad, and the ugly," as Henry Banks, a lifelong resident, put it. He is a former grocery store owner and a respected community leader. His family has lived in the city since the Civil War. We also met with Harry and Shirley Hart at the Second Ward Community Center and then with two ministers, Reverend Tracey Sumner and Pastor Michael Curtis, at the Second Baptist Church across the street.

Our conversations actually focused on the good—that is, on what the community is doing for itself with the city's occasional support. I heard enough explicitly and in the subtexts, however, to understand that the gap between whites and blacks persists. The South End continues to be a separate and distinct place— somewhat like the old East Side. Many older black residents still feel the pain of past prejudice and mistreatment. Some refuse to go downtown at all, and they wouldn't dream of eating at Bun's—the iconic Delaware restaurant and business. The Second Ward white councilwoman representing the South End pays little attention to her black constituents, according to longtime residents. "I've never seen her," Henry Banks said. "I don't know why she hasn't been more visible or in touch." The Second Baptist Church leaders referred several times to "communication barriers" and even now to unpleasant encounters. "There are still places where we are treated differently," Pastor Curtis said.

Harry Hart's story captures well, for me, the South End experience. He grew up there in a close and supportive community. When he was young, his parents protected him from racial prejudice. He didn't know, for example, why he and his friends sat in the balcony at the Strand Theatre, nor did he understand until later why his parents would say, "Don't go in there" about certain downtown businesses. As he grew older, he learned about racism from personal experience and the shocking events reported in the media. "I couldn't believe what I was seeing—fire hoses and dogs turned on people—my people—folks being clubbed by cops on horses, chased, injured, humiliated." (He was referring, I assume, to the three marches on Selma in 1965.)

In high school, he helped organize a school bus boycott because of the way black students were treated on the school buses and in school. "It was time to stand up and protest," he explained. Everyday during the boycott, the students from the South End walked three miles to and from school. White high school students supported them. Ohio Wesleyan students helped them organize. The boycott lasted a few months and ended when the school accepted a list of demands—such as seats on the buses (by the time the buses picked up black students, all of the seats were taken by whites), fairer disciplinary treatment of blacks, and more library materials related to the black experience.

After high school, Harry enlisted in the Marines. His family couldn't afford to send him to college, and he didn't want to work in a factory—about the only local option for young black men. After his military service, Harry lived in California and New Jersey, but he was drawn back to Delaware, initially because his parents still lived there. "But I also wanted my children to have the same sort of growing up experience I had. Delaware was still a place that was peaceful and quiet. I didn't have to worry about putting bars on the windows. Then I became more involved in things—with the growth and the hope that we could still hold on to that small-town atmosphere and also keep up with the times"— and change and improve life for the South End.

Harry and Shirley Hart serve as principal volunteers for the Second Ward Community Initiative (SWCI). She is president of the organization, which serves low-income and high-risk individuals in the Second Ward by providing such activities as youth enrichment; health, wellness, and fitness classes; clinics to enhance young girls' self-esteem; congregate meals for seniors; and youth academic tutoring.

The city has supported the initiative by leasing the old Miller Implement building on Ross Street to the SWCI for one dollar a year. "The town has been very supportive and helpful in getting this community center developed," Harry Hart said as we met in a building slowly being converted from a showroom and garage-like structure into a community center. The city helps with repairs. The organization raises money and seeks donations for program and renovation costs. It was a building dramatically different from the majestic County Courthouse. The place told a story of community solidarity and effort, of Delaware's investment in the South End, and also of difference. The gap remains.

The city provides support, but it can also ignore or marginalize the community. The gas station fiasco is a case in point. The city council did not talk with anyone in the South End about need or desirability before it approved construction of a gas station and convenience store strip mall on London Road. Residents were incredulous. "There's already a gas station across the street," someone would have said. "You're putting it here, near Woodward School, and ignoring the problem of elementary kids walking to and from school? How could you not speak to anyone in the community about its effect?" Well, it was built and operated for a time. But the gas station had closed, and every storefront in the strip mall stood vacant, when I saw it—grass and weeds growing all around it.

"There are decisions being made without our involvement," Reverend Sumner explained. "The leadership in the South End has been here for years, but we either aren't included, or if we show up, people freak out."

The white leadership of Delaware has been redefining and redeveloping the downtown and rebuilding the city's business and industrial base. In an equivalent project and with limited resources, the SWCI and other organizations are redeveloping the South End. Habitat for Humanity has built thirty-five homes in the South End—at least a third for black families—with more planned. The Second Baptist Church was instrumental in starting the Second Ward Community Center. And while Ohio Wesleyan "couldn't give us [the Second Ward Community Center] funding," Harry Hart explained, "it did everything else—gave us furniture and computers. That partnership is alive and well today." For decades the Liberty Community Center served residents broadly. It concentrates now on early childhood education. Each year the residents hold a Community Unity Festival. The South End is acting on its own behalf with energy and commitment. The people there have a powerful sense of place.

Harry Hart and other South Enders want what white Delaware wants—a vital community where individuals and families can live well—can be at home in their corner of the world. I'm impressed by the community spirit and activism in the South End, as I am with the renewal of downtown Delaware

"It has been a good ride here in Delaware," Henry Banks said at the end of our conversation. "I wouldn't live anywhere else. It's a good place to live." Harry Hart said much the same thing, just before we left. "You couldn't drag me out of here now." Both are embedded in the South End much like I was—and in a sense still am—in the North End of Delaware.

I experience Old Delaware as an absence, as well as a powerful presence, in my heart and mind. Now that I am learning about the New Delaware, I feel more and more attached to it. In certain respects, the city seems like the one I knew decades ago. Still, there is a significant difference. It is truly a place of both continuity and change. The handsome, late-nineteenth- and early-twentieth-century building facades downtown, the elegant nineteenth-century brick homes along West Winter Street, the solidity of the university buildings, the homes in the northwest historic district, the East Side, and the South End, the schools, the newspaper—all ground the city significantly and sustain its story through time. The quality of life and the type of community both white and black Delawareans desire seem consistent with Old Delaware—even if the matters of race and class persist in the twenty-first-century city and if the city is much larger now than Old Delaware.

Some of my former classmates, who have lived through the changes in Delaware year by year, experience it as loss. "We used to know everybody when we walked downtown. Now we know almost no one. You used to have people wait on you that knew you. You knew the police. They knew you. You could let your kids go anywhere. Now you don't know anyone." I heard similar sentiments from Harry Humes, a friend of my father and an important community leader: "In those days, when I walked Sandusky Street, I knew about 90 percent of the people. Now I know hardly anyone. Most residents worked in Delaware. Now, many commute to Columbus. We've become more and more a bedroom town." And I remember walking along Sandusky Street with my father. He knew virtually everyone, spoke to them by name, and tipped his hat to the women.

The current leaders look at the situation differently. "The crowds that you see on Friday night, I don't recognize all that many people. People are coming in from other communities in the county," one leader said. Another said, "You have people rediscovering their county seat." A third added, "We know a lot of people around town, but we also don't know a lot of others. That's good. That's wonderful." They celebrate the growth and the new people coming downtown.

Even so, Jack Hilborn expresses some anxiety: "I wonder about some of the newer residents—if they have the same sense of place for Delaware. People move in, work south. It's a bedroom community in some respects. Do they feel the same way [as we do]?" "Yes, they do," the city manager replied and referred to the positive results of a community survey.

New Delaware seems more distributed than old Delaware—more extended and obviously much larger. I might have once said *decentered,* but that would not be accurate. The city is *re-centering* itself. But since it has grown so much and is now home to many newcomers, it is creating a different kind of center. I do sense continuity from past to present, but New Delaware is in fact new. Although I cannot define them—I simply do not know enough—the forms of life for young people now growing up in and around Delaware must be somewhat different from those I internalized. If the new downtown does in fact re-center life in the city, as it seems to be, it will be centering a somewhat different town and not restoring what once was or creating a simulacrum of Old Delaware. The new center will be expressing, enabling, and enacting somewhat different forms of life from those in Old Delaware.

———

I am grounded there in a way that overcomes the relative placelessness of my life and the displacements of modern times: the monoculture of universities, the uniformity of locations across the country, the culture of upward mobility and of moving on, and "the juggernaut of globalization." Without a place, we are in a real sense without a self. We are displaced, even if the peripatetic character and constant movement of modern times seems so natural. We're so possessed by that culture, we no longer know what we have lost.

Over time we might live in many locations and come to know them fairly well, but it is rare—at least in my experience—that we are truly *placed* in those locations, that they shape and define us as our primal and primary places have. We pass through them—having enjoyed and known them to some extent—and then move on, remembering and caring, but not having dwelled fully in them. We learn from each, but it is not the same as being truly placed or truly dwelling in a site or space that has transformed us or that we have transformed into a human place.

I am still writing about "we" and "you." I continue to believe that my desire and need for an identifiable place and for a sense of self, achieved through place, reflects your desire and need. I have made my Whitmanesque turn: "These and all else were the same to me as they are to you."

I have discovered my primal place and the primary centers of my childhood and youth. Delaware and I are tightly intertwined. I am grounded by that place. It enabled me, as I have described, but Delaware also limited and disabled me in significant ways. Even so, my sense of self and sense of place, I now understand, have given me a solid base throughout my life.

I come back, once again, to my sense of place as complex and dynamic. The layers I speak of—or the forces in play—geographic, constructed, psychic, social, cultural, mnemonic, imagined—all intertwine and act to shape individual selves and communities. In play, these elements express, enable, and enact forms of life that influence individuals profoundly. Place is an event—it *takes* place in a real sense.

My academic colleagues might say (with wry smiles) that I am living still according to a small-town master narrative—at least in part. I am living as if life were relatively orderly, coherent, continuous, stable, and secure—when I know that in so many ways it is not. I am living as if people mean what they say and say what they mean—when I know they often do not. So I live—and I believe many of us do—in continual conflict, a small town self, trying to negotiate a complex, uncertain, often dishonest, and violent world.

17.

Oaknoll Farm

Elizabeth Adair Obenshain

I am not placed so fully or embedded so deeply, however, as someone like Elizabeth Obenshain is at Oaknoll Farm in the Virginia mountains. Her homecoming has been complete. After years of professional life away from the farm, she returned in 1994 to the place Sam Obenshain, her father, purchased in 1937, to the house her parents had built in 1941, to the home where she lived from birth to college, and to her childhood bedroom. She lives once again in her primal place.

Her brother, Joe, owns and lives at Blue Ridge Hall in the small crossroads of Mill Creek in Botetourt County, Virginia, in the house where their father was raised and that has been the family home place since the 1840s. During Betty's childhood and youth (she was Betty until she went to college), the Oaknoll Obenshains returned to Mill Creek to celebrate virtually every holiday with the entire family—her grandfather, six aunts, a number of cousins, and others.

The Obenshains' family continuity and identification with place differs from mine and seems uncommon in our modern times—at least for large segments of the population. Beth is so deeply embedded in the land, the farm, her family, and its history that I cannot imagine her leaving unless forced by circumstance. Her story confirms my sense of the power of place and the way self and place can be so tightly intertwined. Her experience also reveals what so many have lost or never known—identification with a specific place, distinct from every other, a place with great character and influence. She is embedded in Oaknoll Farm much as my friends and neighbors in Clover Hollow were placed in their homes and on their farms, and like Jim Cox was in spite of his resistance. He lived a constant interplay of attraction and resistance.

When Beth was growing up, Oaknoll Farm was well outside a small college town, along a narrow, winding country road. Prices Fork Road was so rural the Obenshains could drive cows along it from one farm to another. (The road now is a four-lane, arterial highway where cars travel at forty-five to fifty miles per hour.) The farm covers one hundred thirty acres of rolling terrain and wetlands. It looks toward Brush Mountain to the north. The views across the farm are exceptional. Sam Obenshain was an agronomy professor at Virginia Polytechnic Institute, now Virginia Tech. He studied soils for a living; he acquired land to enjoy, farm, and live on. He was, in a sense, recreating a way of life he'd known growing up at Mill Creek. At the same time, he was pursuing an academic career and a new kind of life.

Sam, his wife, Josephine, their three sons, Dick, Scott, and Joe, and one daughter farmed Oaknoll as a typical mountain family farm. They raised chickens, kept a few hogs, maintained a small dairy herd, raised beef cattle, and grew corn and hay for livestock feed. Beth's mother, a college librarian before her marriage, managed the house and helped manage the farm during her husband's frequent travels around the state and abroad. Beth was milking cows by age four, walking the farm and watching thunderstorms with her grandfather, tagging along behind Frank, the hired man, as he worked. As she grew older, she shared responsibility with her brothers caring for dairy calves, helping with the beef cattle, rounding up cattle for vaccinations and pink eye treatments, and showing dairy cows and beef cattle at the county fair and 4-H shows.

The farm also taught tough lessons. She recalls the time when she and her brother, Joe, "killed" a few dairy cows. They'd herded the cows early one morning into a field of wet clover. The cows gorged, bloated, and several died. That experience and the time her father killed a bull loading it through a chute onto a truck became part of the family's lessons and lore. The truck moved; the bull fell and broke its neck. She'd help raise a calf, come to love it, and then her father would sell it to a feedlot or slaughterhouse. She'd care for a runt pig, keep it alive, and then it would die—just like other small animals she looked after and loved. She learned loss and death early on. Later, she learned to distance herself somewhat from the animals. "I couldn't stand the continual heartbreak."

Beth Obenshain grew up as a farm girl with a deep and permanent attachment to the land. When she and a friend were trying to find the single word that captured each of them, she chose "land." She grew up as a farm girl, yes, but one with a professor as a father, a love of reading from her librarian mother, and an intense interest in school. She learned to farm, but she was also being prepared for an off-the-farm future.

At Oaknoll Beth often passed hours and days by herself. There were no neighborhood children to play with—just the farm, the work, Frank, and her family. "I loved being on my own. I couldn't wait to start a day. I'd get up at four thirty or five, before anyone else was awake, and go outside," listening to early morning birdsongs, waiting on the sunrise, just looking out over the farm. "It was all mine." In warm weather, she'd climb into the hay mow with a book, sit on the bales reading, loving the sweet smell of the hay, listening to the wind and rain as they struck the barn, and gazing out at the oak forest on the other side of Prices Fork Road. (That woods has been replaced by a high-density townhome development.) "I was never scared of being alone." And that sense of self-reliance and confidence, developed in her early years, has sustained her throughout life.

She was raised in a loving and supportive family made even closer by the farm work they shared. "It was a wonderful house. It was a happy house." Beth remembers her mother as a warm, gentle, welcoming woman—a city girl who came to love the farm—who might not even complain when the kids walked into the house with cow manure on their shoes. Her father was patriarchal, but even so, he encouraged Beth and her brothers to make their own decisions about school and life. Beth felt she shared an easier relationship with Papa than he had with his boys—especially Dick and Scott, the two older ones. She was the youngest and the only girl. He demanded less from her.

"I had such a solid base of love and positive identity with my parents. They made me feel like a valuable individual. That sort of thing stays with you. [It] supported me when I went through a divorce. It was an unhappy time, but I never really doubted who I was or felt unworthy."

Her brothers all went to school before she did. She could hardly wait. "I was dying to go to school"—to get on that school bus and go. She could already read, write, and tell time before first grade. At Blacksburg Elementary, she joined other country kids, children from town, and kids from other faculty families. She felt comfortable immediately. All through school, she and her brothers balanced school activities with farm work. They would rush home after school to milk cows and do chores and then return for after-school activities—for Beth, forensics and drama. She felt at home on the farm and in Blacksburg schools—until high school graduation and the prospect of leaving for college.

"The hardest thing for me was going away to college. I felt so sheltered and happy here. I'd never had to redefine myself in Blacksburg. I could [have stayed and gone] on . . . being 'Little Betty Obenshain.'" She knew, however, that she had to leave to become truly independent and mature. Otherwise, "I would always be a small-town, hometown [farm] girl." During her first semester at Westhampton College, she woke up every morning with a stomachache. Near the end of her first Christmas break, she sat alone, late at night by the fireplace, longing to stay but realizing she had to go back.

Beth soon found Westhampton academically deficient and socially "like a women's prison." She even had to ask permission for her brother, Dick, a practicing Richmond attorney, to visit her after seven thirty in the evening. After two years, she and a lifelong friend transferred to the University of North Carolina at Chapel Hill. "Carolina" was a revelation—intellectually rich, socially exciting, and politically active. The university expressed and enacted values that transformed her life.

Beth never rejected her family, Oaknoll Farm, or her family's long Virginia history. She did not have to "wade out" as Jim Cox did from his southern heritage or as, I believe, I had to from Delaware. Her primal place continued to sustain her. A capacious world, however, was opening. She entered it eagerly. She had to leave Oaknoll to become Beth and no longer Betty, except to her family. Jim Cox used to talk about the way a name change expresses or creates a transformation. Walter Whitman, a commonplace journalist, becomes Walt Whitman. Samuel Clemens becomes Mark Twain—both liberated in their new identities. Beth's name change liberated her as well to enter a new world and then return as a mature woman with deep roots at Oaknoll.

After college, Beth pursued a journalism career as a reporter and news editor for more than thirty years—first at the *Richmond Times Dispatch*, then the *Fayetteville Observer*, and finally at *The Roanoke Times*. She relocated from Fayetteville to Lexington, Virginia, when her husband took a position at the Washington and Lee Law School. For several years, she commuted to work at the Roanoke paper. She then moved to Roanoke—closer to the newspaper office and also closer to Oaknoll. When her "dream job" opened—bureau chief at the New River Valley office, in her home county and nearer still to Oaknoll—she leaped at it. She lived first at the home she bought in Blacksburg, but then she returned to the farm to care for her father. Her mother had died in 1992. After her father's death, she became the main managing partner of three family farms. Her homecoming was complete.

Beth Obenshain had graduated Phi Beta Kappa from a major university. She had succeeded professionally. She'd lived in New York City for a year attending Columbia on a journalism fellowship. She had traveled widely in Europe. The farm girl had become quite cosmopolitan. When I told her how I'd described her to my daughters, she responded, "So they'll expect me to step off the plane in bib overalls and Ferragamos?" Intellectually and stylistically sophisticated, she is, nevertheless, still a farm girl. Oaknoll Farm centers her. It remains her primal and primary place.

She left not knowing she would ever return. When she became bureau chief, she never pictured herself living "back on the farm." Circumstances, however, made that both necessary and desirable. It would be too much to say it was her "destiny," but to say less might be equally mistaken. Beth had to leave to fulfill the self bequeathed to her by Sam and Josephine and shaped by Oaknoll Farm. That self developed into a different person while she was away, one who is nevertheless the same person.

"When I came back," she says, "I wanted to live in the different person I'd become since I left"—a woman more socially confident, more attractive, more fully self-reliant, and "with a different political viewpoint." This new self was, nevertheless, still centered on the farm. She was at once the woman who left, but also the woman who stayed emotionally and imaginatively, and then the woman who returned understanding how fundamental Oaknoll is to her life.

"This farm is at the very core of who I am," she said. It is the land itself. "You could blindfold me and put me out on this farm anywhere, and take the blindfold off at night, and I could almost tell you exactly where I was. I know every inch of this farm. I've walked it hundreds of times. I know it like the back of my hand. *It is who I am.*" It is also the house, her parents, and the family. They were and are, in her memory and imagination, all intertwined and constitute the place that has shaped her so profoundly and that has sustained her from the time she left, during decades away, upon her return, and to this moment. Oaknoll Farm "is the rock on which everything else depends. It is a magical place," Beth says.

Like Delaware for me, Oaknoll Farm has changed, yet it is the same place. The farm was once well out of town, but now Blacksburg virtually surrounds it. Look across the four-lane road to the townhouse development and you see residential Blacksburg. Look to the east and you see a new residential village for eight hundred college students. Look to the north across the farm toward Brush Mountain, however, and you see only the farm—the rolling pastures, clusters of trees, wetlands, cattle, the oak forest in the distance, and the mountain. You see a beautiful working farm. You're in the past and the present—almost in a timeless place. Beth Obenshain is embedded in the one but lives in both. She and the farm are still thoroughly intertwined.

———

Through the windows of my study, I see Brush Mountain in the distance—the mountain I crossed almost every day on my way to Clover Hollow. I look across meadows, woods, wetlands, forests, as far as the mountain ridge—land that Beth Obenshain has known since childhood.

I live at Oaknoll Farm now in a new life and splendid, if unexpected, marriage with Beth Obenshain. I've not returned to Clover Hollow in any literal sense, but I do live on a farm, near Brush Mountain, not far from that beautiful valley. I'm satisfying once again the desire I felt first in the woods behind my home at 208 West Lincoln. I am living within the personal culture that has grounded my adult life and been part of my moveable place. I am also living with books and artworks that have been part of the place I carry with me. I am finding Clover Hollow again, through proximity, renewed friendships, and by thinking and writing.

I reside with Beth in a family home that belongs to her, her brothers, and their families. It is not our house together, nor is

Image 14. Oaknoll Farm.

it mine in any ownership sense. Nevertheless, I identify with
the farm and feel oriented here—largely because of Beth and
the modifications we've made to the house. I will never relate to
Oaknoll, however, or to any place as she does here, as Jim Cox did
at Brookside, Caroline Givens Vincel in Clover Hollow, or even
my father in Delaware, Ohio. I've never known anything quite like
their experiences.

Nevertheless, I have discovered my primal place and the
primary centers of my childhood and youth through the experien-
tial and reflective journey I've traveled. I'm grounded in Delaware
in a way that overcomes the relative placelessness of my life. I've
realized my homecoming through memory, research, visits, conversations, reflection, and imagination. I do not in fact live in my
hometown, nor does anyone in my family. Nevertheless, I am both
there and obviously not there. When I visit, the city embraces me

as I reach out to it. I gather it in as the complex, dynamic place it is. When I return to my homes in Virginia or South Florida, I carry Delaware with me as part of my moveable place—this time a place that grounds me in my past and present. For me, Delaware is both abstract and concrete. It is an idea—an image, a story, a meaning. It also exists as an actual place that continues to be part of my life.

I identify with Delaware. I have some sense of its *genius loci*—both the guardian spirit of my youth and the not so different spirit of New Delaware. I feel oriented to the twenty-first-century city, as well as to mid-twentieth-century Delaware. I can walk its streets, drive through neighborhoods, and shop along Sandusky Street now, and in my memory, and know where I am. The streets haven't changed. The downtown buildings retain their solidity. The grand houses along Winter Street still stand. The house at 208 West Lincoln Avenue still serves as someone's residence, as do all the other houses along the street. The South End and the East Side have sustained their distinct identities over time. The university has maintained its fundamental character as it has grown and changed. I am meeting new people whose interests in sustaining Delaware seem similar to the interests of my father's generation.

I can reside at Oaknoll, feel at home, and value the Obenshain family traditions. At the same time, I dwell in or with Delaware, Ohio—the city that made me who I am. It sustains me emotionally, imaginatively, and philosophically. I do indeed dwell there.

Notes and Sources

The Place Is the Thing

Epigraph. Edward S. Casey, *The Fate of Place*, University of California Press, 1998, ix.

The most influential sources for me about space and place are Gaston Bachelard, *The Poetics of Space*, Beacon Press, 1969; Edward S. Casey, *The Fate of Place*, University of California Press, 1998, and *Getting Back Into Place*, Indiana University Press, 2009; Tim Cresswell, *Place: An Introduction*, Wiley Blackwell, 2015; Christian Norberg-Schulz, *Existence, Space, and Architecture*, Praeger, 1971, *Genius Loci: Towards a Phenomenology of Architecture*, Rizzoli, 1980, and *The Concept of Dwelling: On the Way to a Figurative Architecture*, Rizzoli, 1985 (Norberg-Schulz writes about space and place extensively, as well as about architecture); Yi-Fu Tuan, *Space and Place: The Perspective of Experience*, University of Minnesota Press, 2014.

Other helpful sources about architecture include Witold Rybczynski, *The Most Beautiful House in the World*, Penguin Books, 1989, and *Home: A Short History of an Idea*, Penguin Books, 1987; Paul Shepheard, *What Is Architecture?* The MIT Press, 1994; John Brinckerhoff Jackson, *A Sense of Place, A Sense of Time*, Yale University Press, 1994.

I came to Edward Casey's work late. Casey's treatment of place confirms, I believe, my ideas and also gives me two concepts—place as event and place as a gathering or assemblage—which helps explain the way the layers I attribute to place intertwine and act.

Talmage A. Stanley writes well about place in *The Poco Field: An American Story of Place*, University of Illinois Press, 2012, 2, 3: "Place is a social process, the product of human relationships lived out in a specific landscape, in the context of social and cultural forces and conflicts"; and "American middle-class culture, in the main, conveys messages and lessons of moving on. . . . One's place, however central to one's identity it may be, is of secondary importance to social status, economic success, professional advancement, and full access to consumer goods. These lessons of placelessness go to the root of who we are as Americans."

In *Rethinking Home: A Case for Writing Local History*, University of California Press, 2002, 2, Joseph A. Amato writes similarly: "Everywhere, place is being superseded and reshaped. . . . People everywhere live in an increasingly disembodied world, their landscapes and minds increasingly falling under the persuasion and control of abstract agencies and virtual images. . . . Space and time, which once isolated places and assured continuity to experience and intensity to face-to-face interaction, have been penetrated, segmented, and diminished by surrounding forces and words. The coordinates of community, place, and time no longer define identities."

I have been reading about self and identity virtually all of my adult life. Both self and identity are themes in my books, *Walt Whitman: The Drama of Identity*, Michigan State University Press, 1973, and *Loren Eiseley: The Development of a Writer*, University of Illinois Press, 1984.

The matter of memory, history, and the past is central to my essay "The Past in the Present: The Greater Rural Newport Historic District," *Appalachian Journal*, Fall 2004.

1. James Melville Cox and Brookside Farm

My introduction is based on extensive interviews with Jim and Marguerite Cox; on a National Register of Historic Places Registration Form: "Brookside Farm & Mill, Grayson County, Virginia," OMB No. 1024-0018; on "Cox Baseline Documentation Report, Grayson County, November 16, 2006," by the Virginia Outdoors Foundation; as well as on the architectural and historical sources I cite below.

Page 6. Bachelard, 4, 7, xxxii; Norberg-Schulz, *Genius Loci,* 6.

Page 12. Seamus Heaney, "A Herbal," *Human Chain*, Faber and Faber, 2010, 43.

2. Placeless in America

Page 14. Talmage Stanley. See reference above.

3. Clover Hollow: Our Sanctuary

My own publications deal with Clover Hollow and Newport: "Mark Givens: The Last Full-Time Farmer in Clover Hollow," *Now & Then*, Summer 1999; "Fitting In: Outsiders in a Rural Mountain Community," *Now & Then*, Spring 2006; "The Woman Who Stayed and the Man Who Left," *Appalachian Heritage*, Winter 2000; "The 1892 Givens Home-place: The Fate of a Mountain Farm," *Appalachian Heritage*, Fall 2001; "Insiders, Outsiders, and the Struggle for Community," *Appalachian Journal*, Spring 1999; "The Past in the Present,"

Appalachian Journal, Fall 2004. The first *Appalachian Journal* piece deals with the power line fight; the second with the historic district (see pp. 74–75).

The historical information in this chapter and in chapter 5 is drawn from Hattie Miller, *A Story of Newport and Its People,* published privately, 1974; Susie Reid Keister, "A History of Newport, Virginia," *The Virginian-Leader,* Pearisburg, Virginia, 1969; Dorothy Hall Givens, *A Givens-Hall Family History from Pre-Revolutionary Times to 1970,* Radford, Virginia, 1971, and *The Givens Tree Spreads,* Radford, Virginia, 1982; *Giles County, Virginia: History—Families,* vols. I and II, Giles County Historical Society, 1982, 1994 ; Douglas D. and Perry D. Martin, "Newport, Virginia: A Crossroads Village," *The Smithfield Review,* vol. 1, 1997; an extensive archive maintained by Nancy Kate Givens; and many hours of interviews.

On rural architecture: Henry Glassie, *Vernacular Architecture,* Indiana University Press, 2000 and *Folk Housing in Middle Virginia: A Structural Analysis of Historical Artifacts,* Knoxville, 1975; "Eighteenth-Century Cultural Process in Delaware Folk Building" in *Common Places: Readings in American Vernacular Architecture,* Dell Upton and Michael Vlach, editors, University of Georgia Press, 1986; Fred B. Kniffer, "Folk Housing: Key to Diffusion" in Upton and Vlach; K. Edward Lay, *An Architectural History of Albemarle County, Virginia,* Charlottesville, 1997; D. H. Jacques, *The House: A Pocket Manual of Rural Architecture, or, How to Build Country Houses and Out-buildings,* New York, 1859.

4. Three Meadow Mountain: Homage and Innovation

This chapter is informed by interviews and the cited architectural sources. Norberg-Schulz's observation about transforming space is especially instructive. The house, in his terms, has transformed an open meadow and mountainside—open space—into a human and natural place. It has uncovered some of the potential meanings of the space. "The existential purpose of building (architecture) is therefore to make a site become a place, that is, to uncover the meanings potentially present in the given environment." *Genius Loci: Toward a Phenomenology of Architecture,* Rizzoli: New York, 1980, 18.

Page 27. Witold Rybczynski, *The Most Beautiful House in the World,* Penguin Books, 1989, 93.

Page 29. Henry Glassie, *Vernacular Architecture,* Indiana University Press, 2000, 117.

Page 33. "Propriety" and "decorum" are Rybczynski's terms.

Page 34. Paul Shepheard, *What Is Architecture?* The MIT Press, 1994, 105–6.

Page 35. Christian Norberg-Schulz, *Existence, Space, and Architecture,* 27, 39, 114.

5. Clover Hollow: The Place

See the citations above for Chapter 3.

6. The 1875 Lafon Home Place

7. The 1892 Givens Home Place

Chapters 6 and 7 are based on extensive interviews and the architectural sources I have cited earlier. Parts of chapter 7 are adapted and expanded from my essay, "The 1892 Givens Home Place: The Fate of a Mountain Farm," published in *Appalachian Heritage*, Fall 2001. Permission granted by *Appalachian Heritage*.

Page 63. Norberg-Schulz, *Genius Loci*, 6.

Page 63. Casey, *Getting Back Into Place*, xv.

Page 64. Norberg-Schulz, "Two Aspects of Dwelling," *The Concept of Dwelling*, Rizzoli, 1985, 15ff.

8. Outsiders Fitting In

Parts of this chapter are adapted from my article "Fitting In: Outsiders in a Rural Mountain Community," published in *Now & Then: The Appalachian Magazine*, vol. 22, no. 1. "We observe First NA Serial Rights, the right to republish rests with the author." Randy Sanders, managing editor, *Now & Then*.

Page 67. The concept of a hand reaching out comes from John Brinckerhoff Jackson, "Working at Home," *A Sense of Place, A Sense of Time*, Yale University Press, 1994, 145.

Page 67ff. My account of Mark Givens is adapted from my article "Mark Givens: The Last Full-time Farmer in Clover Hollow," published in *Now & Then*, vol. 16, no. 2.

9. Hometown Delaware

This part of *Hollow and Home* is based on interviews, Delaware County Library and Delaware Historical Society archives, public school records, high school yearbooks, *The Delaware Gazette,* Columbus, Ohio, newspapers, and my own memories. In March 2013, I lectured on Hometown Delaware at Ohio Wesleyan University as part of the annual Vogel Delaware History Series.

10. A Boy from Columbus. A Man *of* Delaware, Ohio

This is based on extensive interviews and research in the following: twenty-three years of *The Delaware Gazette*; four years of the *Ohio Wesleyan Transcript*; issues of the *Columbus Dispatch, Columbus Citizen,* and the *Ohio State Journal*; Ohio Historical Society, Ohio Wesleyan University, and Columbus South High School archives; public school records in Delaware, Ohio; high school and college annuals; and on my privately published memoir, *Searching for Ervin,* 2006.

11. 208 West Lincoln Avenue

12. Delaware City Schools

13. Downtown Delaware

Page 101. Akiko Busch, "Preface," *Geography of Home*, Princeton Architectural Press, 1999.

Page 105. Rybczynski, *Home*, 231.

Page 105. Gaston Bachelard, *The Poetics of Space*, 15, xxxii, 7.

Page 121. Robert Frost, "The Black Cottage," *Frost: Collected Poems*, Library of America, 1995, 60. W. E. B. Du Bois, "Of the Dawn of Freedom," *The Souls of Black Folks*, Library of America, 1986, 372.

14.The Road Out: Ohio Wesleyan University

This draws on the *Ohio Wesleyan Transcript*, the *Ohio Wesleyan Magazine*, Ohio Wesleyan University Library archives, interviews with classmates and fraternity brothers, and my own experiences at Ohio Wesleyan.

Page 148. Bachelard, xxxii.

Page 149. Walt Whitman, *Song of Myself*, section 1.

15. A Moveable Place

The Whitman quotations are from "Whoever You Are Holding Me Now in Hand" and "Song of Myself."

Page 155. Edward S. Casey, *Getting Back into Place*, 275.

Page 156. "Culture means to transform the given 'forces' into meanings which may be moved to another place. Culture is therefore based on abstraction and concretization. By means of culture man gets rooted in reality, at the same time he is freed from complete dependence on a particular situation": Norberg-Schulz, *Genius Loci*, 170.

16. New Delaware: The Place Is Still the Thing

"New Delaware" is based on my previous research and long experience. It also draws on (1) A long conversation with Delawareans R. Thomas Homan, city manager; Jeff Benton, county commissioner; Jack Hilborn, insurance executive; Roger Koch, architect and developer; Ana M. Babiasz, president and CEO of Fidelity Federal; Brent Carson, board president, Delaware County Historical Society; and (2) conversations about the East Side and South End with Mary Jane Santos, Henry Banks, Harry and Shirley Hart, Reverend Tracey Sumner, and Pastor Michael Curtis.

Helpful sources about small towns are Paul L. Knox and Heike Mayer, *Small Town Sustainability*, Birkhäuser Verlag, 2009; Richard O. Davies, *Main Street Blues: The Decline of Small-Town America*, Ohio State University Press, 1998; Richard C. Longworth, *Caught in the Middle: America's Heartland in the Age of Globalism*, Bloomsbury, 2008.

Page 171. Habitat for Humanity information from Todd C. Miller, executive director, Delaware County Habitat for Humanity.

Page 173. "Juggernaut of globalization" is from Edward S. Casey, *Getting Back into Place*, xxxv, as is the notion of *taking* place.

17. Oaknoll Farm: Elizabeth Adair Obenshain

Based on an interview with Elizabeth Obenshain.

Index

www.ingramcontent.com/pod-product-compliance
Lightning Source LLC
Chambersburg PA
CBHW071741270326
41928CB00013B/2763